WRITING THE WEST COAST

Writing the West Coast

IN LOVE WITH PLACE

EDITED BY
CHRISTINE LOWTHER
& ANITA SINNER

RONSDALE PRESS

WRITING THE WEST COAST

RONSDALE PRESS
3350 West 21st Avenue
Vancouver, B.C., Canada
V6S 1G7

Typesetting: Julie Cochrane, in New Baskerville 11 pt on 15
Cover Design: Julie Cochrane
Cover Art: *The Eik Cedar*, a painting by Joanna Streetly; photo credit, Wayne Barnes
Paper: Rolland Enviro Cream (100% recycled)

Ronsdale Press wishes to thank the following for their support of its publishing program: the Canada Council for the Arts, the Government of Canada through the Book Publishing Industry Development Program (BPIDP), and the Province of British Columbia through the Book Publishing Tax Credit Program and the British Columbia Arts Council.

Library and Archives Canada Cataloguing in Publication

Writing the West Coast: in love with place / edited by Christine Lowther and Anita Sinner.

Includes bibliographical references.
ISBN 978-1-55380-055-2

1. Queen Charlotte Islands (B.C.) — Biography. 2. Clayoquot Sound Region (B.C.) — Biography. 3. Ecology — British Columbia — Queen Charlotte Islands. 4. Ecology — British Columbia — Clayoquot Sound Region.
I. Lowther, Christine, 1967– II. Sinner, Anita, 1967–

FC3845.P2A23 2008 971.1'12 C2008-901436-7

At Ronsdale Press we are committed to protecting the environment. To this end we are working with Markets Initiative (www.oldgrowthfree.com) and printers to phase out our use of paper produced from ancient forests. This book is one step towards that goal.

Printed in Canada by Marquis Book Printing, Quebec, Canada

For the record, I am helplessly in love
with Earth as my home. I am completely satisfied
to be here. My allegiance is given. I do
not need or want another heaven.

— SUSAN KAMMERZELL

ACKNOWLEDGEMENTS

We are particularly grateful to Ronald Hatch of Ronsdale Press for his commitment to this project and his unwavering creative spirit. The process has been a joyful one because of Ron. His mentorship and his willingness to showcase the talents of new writers is truly exceptional.

We extend our special thanks to the many authors who have made this book possible, for their dedication to the places called home, their willingness to share private moments, and their patience during the process.

To our families and friends who have offered their encouragement, advice, knowledge, errands and other favours, thank you (and to any we have forgotten) — in somewhat random order: Jen Pukonen, Warren Rudd, Julie Cochrane, Mike Yeager, Clayoquot Writers' Group, Kate Braid, Sharon Butala, Sherry Merk, Maryjka Mychajlowycz, Christine Wiesenthal, Joanna Streetly, Jackie Windh, Marcel Theriault, Jan McDougall, Adrienne Mason and Bob Hanson, Adrian Dorst, Barb Campbell and Linda White from Tofino Library, Keith Harrison, Kevin Bruce and Naomi Carson, Alex and Dianne Nikolic, Norma Dryden, David Griffiths, Jackie Prescott, Hiro Boga, Brittany Smith, Beth Lowther, Shirley Langer, Chris and Rowan Jang, Rick Barham, Sandor Csepregi, Cameron Dennison, Harvey McGilvery, Bernadette Mcallister, Sharon Doobenen, Mark Spoljaric, Josie Osborne, Wayne Weins, Andrea Lebowitz, Fourth Street Natural Market, Tofino's book shops, Raincoast Books, New Star Press, Oolichan Books, Orca Books, and *Sea Kayaker* magazine. Thanks also to those who submitted essays that we could not include in the project.

EDITORS' NOTE ON FIRST NATIONS PLACE NAMES: *While it is common to see traditional villages spelled with a silent "h" — for example: Hesquiaht, Opitsaht, Ahousaht — these words actually mean the people from these places. In this book we recognize first nations spelling of their homes: Hesquiat, Opitsat, Ahousat, while acknowledging that different authors spell these place names both ways.*

CONTENTS

Introduction / 1

- Arriving -

BRIAN BRETT
The Beaches of Clayoquot / 11

ALEXANDRA MORTON
Beloved / 17

DARCY DOBELL
Echosystem / 23

NADINE CROOKES (KLIIAHTAH)
Being Nuu-Chah-Nulth / 29

JOANNA STREETLY
Finding Home / 33

CHANDRA WONG
Uncharacteristically Simple / 41

HELEN CLAY
Called Home / 45

- Yearning -

ANDREW STRUTHERS
Excerpts from *The Last Voyage of the Loch Ryan* / 53

CHRISTINE LOWTHER
Facing the Mountain / 60

MICHAEL SCOTT CURNES
All in a Day's Dream / 73

KEVEN DREWS
Never Say Never / 82

JOANNA STREETLY
Midnight at Catface / 89

SHERRY MERK
Love Song to Clayoquot Sound / 91

- Immersing -

SUSAN MUSGRAVE
An Excluded Sort of Place / 101

DAVID PITT-BROOKE
November Day of the Dead: Seeking Y'aq-wii-itq-quu?as (Those Who Were Here Before) / 107

ELI ENNS
From the Heart of Clayoquot Sound / 113

BETTY SHIVER KRAWCZYK
Excerpts from *Clayoquot: The Sound of My Heart* / 123

VALERIE LANGER
A Groundswell is a Wave / 132

DIANNE IGNACE
Thirty Years in Hesquiat / 142

ROB LIBOIRON
A Sound Existence / 150

- Lingering -

ADRIENNE MASON
Stone Heart / 159

KATE BRAID
A Series of Poems on the Paintings of Emily Carr / 168

KEITH HARRISON
Hornby Island: The Nature of Home / 174

GREG BLANCHETTE
Forty Kilometres from Home / 179

JANIS MCDOUGALL
So, When Are You Moving Back? / 189

ANITA SINNER
The Sensual Coast: Living in the Everyday / 195

- Encountering -

BRIONY PENN
Sex in the City: Love in the Forest / 205

CAROLYN REDL
On a Quest for the Western Screech Owl / 210

BONNY GLAMBECK
Commuting by Kayak / 220

FRANK HARPER
Bright Solstice Darkness / 227

JOANNA L. ROBINSON AND DAVID B. TINDALL
Defending the Forest: Chronicles of Protest at Clayoquot Sound / 232

HANNE LORE
Contact Luna / 248

CATHERINE LEBREDT
A Collection / 254

Contributor Biographies / 265

INTRODUCTION

The great land mass of North America meets the northern Pacific Ocean in a jagged array of mountains, deep valleys, fiords and islands. Clayoquot Sound on the central west coast of Vancouver Island is home to the largest intact temperate rainforest remaining on the Island. Here, where great winds and ocean swells bring fog and rain that can last for weeks, ecosystems of unparalleled beauty and diversity have evolved.

Writing the West Coast explores living on this western edge and, by extension, represents paths to awareness, understanding and being in one's bioregion. Perhaps real love comes from the day when we glean our information from our surroundings rather than from computer screens, although we have both. Incoming weather systems, moon phases and tides, wildlife behaviour, even how high or low (fast or slow) a creek is running can inform our daily lives. This kind of connection, ingrained in first nations culture —

learning to read the natural signs around us, and realizing how to use that information — makes us feel at home in nature. It fosters appreciation of and gratitude toward the land. If our link to place goes back generations, we feel in our bones that we belong. Conversely, if we come here fleeing alienation and then find solace in beauty, our affection is quickly gained. Further, when we experience coastal storms and their shrinking of our own significance, our respect is increased, our "place" as mortal humans acknowledged and confirmed.

A common story among non-aboriginal inhabitants is that they came to visit, and decided to stay. They were captured by a beautiful area, by a magnetic, intriguing environment, by invigorating air stinging with salt — and were transformed. As visitors, they found it relaxing and rejuvenating. But as home the west coast is not an easy place. In defence of the ancient rainforest, so much of which has been destroyed by human rapacity, the inhabitants of Clayoquot have placed their own bodies between forest and chainsaw. In 1993, in the largest civil disobedience action seen in Canada, the inhabitants of the Sound were joined by thousands of protestors from other parts of the world.

Clayoquot is both Nuu-chah-nulth traditional territory and a UN–designated Biosphere Reserve, offering such beauties as Pretty Girl Cove, the Ursus Valley, the Megin River, the Moyeha watershed, Meares Island, the Clayoquot River Valley, the Sydney Estuary and Flores Island. Some of these areas are protected from industries such as logging, Atlantic salmon farming and mining. Some are partially safeguarded while some remain completely unprotected. Often the protected areas are fragmented and vulnerable to roads, which can be built through them to reach unprotected areas. Indeed, anyone flying over certain parts of the Sound today will hear chainsaws, roadbuilding blasts, and ancient trees falling. Granted, the volume of trees cut in 2007 was one third of 1993's volume, but the forest industry's appetite for "fibre" continually threatens the last remaining intact valleys of the region.

Although the tourist brochures like to speak of the area as "wilderness," in fact the native villages have been here thousands of years. Across the harbour from Tofino is Opitsat, Esowista is at Long Beach, Ahousat is on Flores, and Hot Springs Cove has its own village. There are many more traditional sites such as Echachist, Yarksis, Kakawis and Hesquiat. From ancient middens and culturally modified trees, to a two-hundred-year-old bead-encrusted anchor dredged from the deep, and on to our present-day villages and towns, humans have created links to the coast, altered the area substantially, and have themselves been changed.

For visitors, one of the more popular sites is the Meares Island Big Tree Trail. Here a red cedar, "the Hanging Garden Tree," estimated to be a thousand years old, impresses visitors with its massive diameter of 5.9 metres. Before the "Earth Mother" cedar on Meares fell in the autumn storms of 2006, it was Canada's largest tree by volume, at 293 cubic metres, and approximately 1,500 years old. But the popularity of large trees can cause people to overlook the fact that the rainforest comprises a wide range of vegetation. In addition to western red cedar, at the lower levels one finds western hemlock, sitka spruce, amabalis fir, shore pine and red alder. Douglas fir is much less common. An understorey of salal, huckleberry, salmonberry, fungi, nurse logs and hundreds of species of ferns nurture new tree seedlings. At the higher elevations one finds yellow cedar, mountain hemlock and balsam.

As a result of the media exposure given to the extraordinary beauty of the area in recent years, Tofino now receives upwards of a million tourists each year, concentrated in the summer months, when there is a chance that the area's three and a half metres of annual rainfall might ease up. Travellers come to see the big trees, walk and surf the miles of beaches, watch whales, try kayaking and soak in the hot springs. Accommodating so many visitors is a constant challenge — so much so that even with the rainfall, water shortages have become an issue. Global warming was the buzz in Tofino when the town shut down on its busiest weekend of the year, Labour Day of 2006, for lack of water. In addition, the Esowista native village has periodically lacked clean water. Housing for locals is another heated issue. Visitors are the priority; they are our income. But residents serve them, and have a right to secure and affordable housing. As Tofino lies at the end of the road on a narrow peninsula, however, room is running out.

Our initial idea for this collection was to compile a celebration of nature-writing focused on Clayoquot Sound. The submissions that came in were more complicated, however — with mosaics of memoir, humour, nature, research, life writing and activism — all inspired by Clayoquot and other regions of the west coast. Locales south of Clayoquot included here are East Sooke near Victoria and Hornby Island in the Strait of Georgia, where the climate is slightly drier and one finds arbutus trees and threatened Garry oak ecosystems, as well as plants like camas and cacti. To the north of Clayoquot, Nootka Sound is featured, where Captain Cook landed in 1778, at what he named Friendly Cove. Also to the north of Clayoquot is Haida Gwaii, or the Queen Charlotte Islands. The region here is similar to Clayoquot Sound in mild temperatures, heavy rainfall, rainforest, upland bog, salmon streams, estuaries and kelp beds. Finally, there is the Broughton Archipelago, which lies roughly between Port McNeil on the northeastern edge of Vancouver Island and the British Columbia mainland. The marine provincial park here contains one of the most under-represented terrestrial ecosystems in the province — the Outer Fiordland Ecosection Coastal Western Hemlock very wet maritime submontane variant. Currently only 1.3 percent of this ecosystem is protected in B.C.

It is our love for these acutely western places in their changing moods that keeps us all here, and urges us to share our experiences in writing. This anthology quickly became an intimate collection about island life, a vantage point outside of both the metropolitan and the rural — offering a third viewpoint, the far west of the west coast. Here we find lives of contrast: people alone or in a small village within vast wild spaces. Such lives are often lived between the rainforest and rugged rocky headlands, between the peaceful sound of incoming waves and the sudden rogue wave, between deep silence and terrible storms. The breadth of writing in this collection, from renderings of beauty to profound insights into the struggles that shape lives, demonstrates the many ways we come to be in love with place.

To be in love with place can also mean to be in tension with place. Some authors describe their bond with nature as unrequited love, paralleling the loss or absence of home due to lack of affordable housing or appropriate work, as if our cherished ideal location does not want us. This love can be sobered by grief as one watches a place change with development and tourism, or as one feels the date of departure loom closer. Love can be painful. For some authors, it can cause fear as the effects of environmental degradation call for an ever deeper commitment. Such relationships to home require more than a passive, complacent love. Our dedication can actually bring resentment when we come to a place for a deeper sense of community and end up lonely. This loneliness, nevertheless, can reshape identity.

From the lived experiences shared by the writers in this collection, five key approaches or themes emerged: arriving, yearning, immersing, lingering, encountering.

Arriving to find home includes writing that describes coming to a destination in life through place, attaining a state of peacefulness, or for some, a rebirth. Arriving is a pronouncement, an act of conveying happenings which interface between then and now, defined by social and cultural shifts in Canadian society. Narratives indicting the political miscues that reshape the west coast — in particular, the region of Tofino — come together with stories of personal and professional challenges, of science and artful living, and rich versus poor. Here we are reminded of the importance of attending to the constructs that shape our lives in times of increasing disconnect from one another, and from nature. Brian Brett reflects soberly on possible futures for Clayoquot, thinking back to the wild times he spent here in the sixties. Alexandra Morton explores her fierce love of and inability to abandon the Broughton Archipelago with its disappearing pink salmon. Darcy Dobell asserts that an awareness of natural forces stirs an internal compass that orients us to the world. Nadine Crookes describes growing up Nuu-chah-nulth and her gratitude toward her elders. Joanna

Streetly remembers fleeing from home only to find it, unexpectedly, kayaking in her newly discovered Canada. Chandra Wong discusses creative alternatives to the housing shortage. Helen Clay recounts her signpost-filled journey from Exmoor to Clayoquot.

Yearning for the peace and beauty we call nature is written with longing, compassion and desire expressed as milestone experiences in special places. These stories are filled with tenderness, and tell of writers deeply moved. From anecdotes of eccentric dock-dwellers to questions of the cosmos, chronicles of illness and the role of nature in recovery, writers bring forward perspectives of caring and attend to our moral responsibility to nature. Andrew Struthers gives us memoirs of an unconventional life and his sense of humour. Christine Lowther shares wildlife encounters from her floathouse. Michael Curnes dreams of returning to his old haunts. Keven Drews bravely offers his account of healing-by-surfing. Joanna Streetly encapsulates a transcendent moment atop Lone Cone in the moonlight. Sherry Merk reveals how, as a single mother, she worked so hard to stay in Tofino that, after ten years, her health broke down and she had to leave her beloved landscape and community.

Immersing oneself in natural and wild surroundings presents narratives of embedding the whole person in nature-scapes: a practice of meaning-making that requires merging self with nature. Richly textured accounts take readers on walks in Clayoquot or through the daily round on Haida Gwaii. We witness history from the point of view of a young first nations man, and share in the lives of feisty forest protectors and resilient modern-day pioneers. Susan Musgrave shows why she prefers "an excluded sort of place." David Pitt-Brooke investigates and enjoys an ancient hidden midden site. Eli Enns announces the many ways his people, the Tla-o-qui-aht, are improving their lives. He also suggests highly localized, more traditional values in tree-harvesting. Grandmother-activist Betty Krawczyk describes living in a remote A-frame under mountains that have been clear-cut, and how this leads her to join the blockades. Valerie Langer compares Canadian rainforests to remnant European ecosystems — and faces a tsunami warning.

Dianne Ignace portrays a remote lifestyle completely alien to the dominant urban mindscape. Similarly, Rob Liboiron contemplates the advantages and dangers of dwelling off the beaten track.

Lingering in spaces of contemplation allows writers a meditative engagement as they describe in detail their experiences of abiding by the rhythms of the world around them. Stories of lingering involve passing, staying, hesitating, waiting, with a continued presence in nature-scapes. These are chronicles of living in union with the local geography, feeling the contours of the physical world and conveying the quality and grain of what it is to touch ephemeral experiences. Adrienne Mason enacts leaps of faith in juggling family life with career pastimes such as the catching, weighing and banding of murrelet chicks. Kate Braid responds to paintings by Emily Carr, the Victoria artist who first introduced many of us to the imagery of the west coast. Keith Harrison takes us along for a day's meander on Hornby Island. Greg Blanchette tells a story of loneliness. After thirty years in one place, Janis McDougall proves to her city friends that she is too deeply in love ever to go back. Anita Sinner acknowledges the privilege that solitude allows.

Encountering the natural world in diverse and inspiring ways, writers share chance or unexpected meetings. Such experiences may be shaped by difficulty, ease, silence, exuberance, active participation or passive observation, but always they are written as coming to nature. With her usual humour — but seriously — Briony Penn critiques *Sex and the City* while praising love in the forests. Bonny Glambeck illustrates the joys and woes of commuting by kayak. A simple tale of being thwarted by the elements is presented by the late Frank Harper. Carolyn Redl is cajoled down logging roads in search of a western screech owl. Joanna Robinson and David Tindall provide their findings in a survey of blockade arrestees. Their anthropological study of the Clayoquot protests brings our attention full-circle, to community concerns for preservation and sustainability, as well as a discussion of economic, social and environmental solutions. Hanne Lore guides us to Nootka Sound where we see the stray orca, Luna, as Tsuxiit, through the eyes of the Moachaht people. And we are thrilled to include writings by the late Catherine Lebredt, who maintained a profound reverence for the natural world. She observes wolves and raises an orphaned seal pup.

All authors in this collection, whether new or established, contribute to creating a community of writers passionately engaged in searching for home, the heart's hub, where we find purpose and a sense of belonging in creative, contemplative and aesthetically revealing ways that are uniquely west coast.

As we invite you to enter these pages, we are mindful that, because of the mass arrests during the logging blockades in 1993, it is often thought that everything is finally settled in the woods. But most of the coast, including much of Clayoquot Sound, has been logged or remains unprotected, and there are more and more controversies concerning the appropriate use of resources in these waters and on these shores. In this age of climate change, it is notable that the rainforest holds thousands of years' worth of stored carbon dioxide, and Clayoquot is one example among many that warrant greater attention. Indeed, parts of the Sound are being logged by various companies as this book goes to print. It has been important to us to produce this book on Ancient Forest Friendly paper. We know nothing is ever certain in our changing world, but perhaps, in our own ways, wherever we are, we can each find a moment to preserve not only the specialness that is home — as in our case, the west coast — but find ways in our everyday lives to make changes, to redress the ways of life that put all the world's paradises at risk.

— Christine Lowther
& Anita Sinner
February 2008

Arriving

The Beaches of Clayoquot

– BRIAN BRETT –

We arrived at dusk and didn't know where we were. I thought it was Florencia Bay . . . Wreck Beach. . . . We descended a muddy, scrub-wood bluff eroding onto the beach. More like a cliff, it seemed forty feet straight down, and it was a struggle to reach the high tide line where we camped. The grey plain of the ocean washed against the distant shore, barnacled boulders surfacing like decaying skulls out of the sand. It was haunting, that sand, made eerie by the ocean's conversation with itself — the patterns a rippling code, almost decipherable, but not quite.

There was nothing between us and Siberia except water.

It was beautiful but dangerous. Wary that the incoming tide might cause trouble in the night we built bonfires in a line way out onto the beach; we were young and enthusiastic and we hauled tremendous logs with the eyes of birds and the fins of whales out onto the sand, thinking that if the flames were doused by the tide, we would know how dangerous this beach was becoming.

The fires were enormous, glowing beacons on the black sand, like memories frozen into the synapses of the brain. I don't know where we found all the driftwood, but in the frenzy of our youth we heaped up these fires. Memory. Fires in the night.

Years earlier, I found myself standing beside a gravel road. I was young and lonely, long-haired, wearing snood boots and a poncho over my jeans and jean jacket. I knew where the gravel road went, but I was still lost. A milk truck picked me up, the jugs clinking in their cases as the truck bounced up and down the narrow hills and switchbacks overlooking what seemed an endless inlet.

It was the sixties and I was determined to go nowhere.

The great swath of Long Beach was empty when I bushwhacked my way through an overgrown trail that was barely visible where the truck driver to Tofino dropped me off. I pulled together a shanty of driftwood and made myself a shelter. The next morning, awoken by the thump of ravens on the roof of my shelter, I dropped a tab of LSD and spent the day staring at the wrinkles in the sand — wondering if I wanted to stay alive. I never saw another person for three days.

This I thought, in my childish acid-enhanced naivety was the beginning of a changed life.

Then we started drinking. When the first beacon went out I cheered at the confrontation between fire and water, the eternal war. It seemed like there was nothing but war, beautiful war, terrible war. War everywhere — war internal and war in the jungles of Asia. The war against the trees — the eradication of old growth forests around the world. Watching those fires made me think of the glowing rivers of napalm, how they poured it down on the skin of villagers as they fled in terror.

The next time I returned to Long Beach, only a couple of years later, I entered the modern circus. Windsurfers on wheels whistled down the beach, and cars dodged families and children and hippies before slamming into mud holes concealed by a slick of water, where they sank like wicked colonialists trapped in quicksand in a Bomba movie. The police were chasing a drug dealer and they too sank into a sand pool. The car could have been rescued. No one helped them, they were too busy cheering as the tide came in, pounding the car out of sight.

There was dog shit everywhere. Garbage and plastic bags and tarps littered the driftwood line, and toilet paper hung in the trees. Endless parties. Stoned-out hippies. It was a carnival on sand. So I worked my way up to the next beach and rediscovered the lonely magic of a coast that didn't need LSD. This time, still suicidal, I found my way around a point and encountered a magic cave. Sitting in the cave, weeping and mourning a troubled childhood too difficult to recount, I was determined to wait until high tide sealed the entrance. Then a blast of light stabbed through a tiny crack in the cave's roof, and my world filled with glitter, sea anemone, urchins, chitons and crabs, all the gaudy tidal animals glistening with life. Once my heart was back in my throat, I fled the cave and fought my way through the surging tide, back to the beach, renewed once again.

The next fire went out more quickly, dramatically, and we howled a whoop of delight as Pacific's mighty wave humped down on the flame in the darkness. We passed our glass-handled gallon of cheap red wine around more often. This was wilderness. This was an adventure.

I arrived at the sign outlining all the instructions and warnings and regulations, and pulled my van off the paved highway, into the designated parking lot. Sleeping on the beach was illegal, and campers were relegated to a standard federal government campground layout, where each individual could camp on hard-packed gravel next to a neighbour who was playing an ear-splitting ghetto blaster. You couldn't see the beach, but you could hear it when the ghetto blaster was off. You couldn't pound a tent peg into the hard-packed gravel.

We found a hotel with several lodges and a dozen cabins. We rented one of the cabins. They were so close together I could nearly touch the other cabins on each side of ours. I thought, how strange that we would leave our own home, where we couldn't see a neighbour's house, to pay for an expensive cabin in a crowded resort.

We moved the noisy fridge into the bedroom and the bed into the front room. Then I pulled the curtains closed to about a fifteen-inch gap. Early in the morning, or sometimes at supper hour, when no one was on the beach for a few minutes, we would lie on the bed in our darkened cabin, and I would stare through the gap in the curtains and remember this wild coast of my youth.

As the remaining fires slowly went out, one by one, we became so intoxicated we were cheering with a kind of madness that, in retrospect, had a scary edge — celebrating the extinguishing of each fire with another drink.

The night began to rain big, heavy drops upon us. A storm was approaching. I could feel the air change against my face even as I grew more insensible.

Later, I woke up in my tent, half drunk and half hung-over, my feet wet and an egg and a loaf of bread floating past my head in the night only faintly illuminated by the last fire, the one behind us, near the bluff. We were in the middle of a coast version of a hurricane, the tide raging, our gear ruined.

Some time ago I heard an interview with Canada's great poet P.K. Page. In her late eighties now, she's as vibrant and sharp as fifty years ago. The interviewer, in one of those brilliant moments of radio, suddenly changed her line of questioning and said: "Is it all over?"

This took Page aback. There was a brief silence. "Yes," she said, in her dignified voice. "I would have never dreamed that I'd say this five years ago, but we have gone too far."

Although this is a conversation rewritten by my unreliable memory, that's the gist. It was a shock to realize one of Canada's finest poets, with a gift for romance and dreams, had come to the conclusion our planet has gone beyond the point of no return.

Her simple statement was a shock that stayed with me. It is said that George Orwell once remarked: "In a time of universal deceit, telling the truth is a revolutionary act."

We struggled up the wind-blasted muddy cliff in the night, clinging to branches, hanging snags, and rocks eroding out of the bluff, fearing for our lives. Then we stripped in the dark and wrung out our mud-crusted,

Catface Mountain, Clayoquot Sound. (PHOTO: JEN PUKONEN)

Chesterman Beach, Tofino: a glimpse of sun. (PHOTO: JEN PUKONEN)

sopping clothing, feeling doomed, cursing our idiocy, raging like three mad King Lears against the storm.

This was more than thirty years ago, in the last, crazy years of my youth, and I don't know if I want to return to Clayoquot. Sometimes, memory is more important than reality.

A good chunk of Long Beach is park, preserved in a domesticated kind of way. It even has an indoor, heated theatre for campers. The beach with the magic cave is now wall-to-wall houses for the wealthy and celebrities — movie stars and singers, the taxes so high many of the original settlers have been forced to sell. The cave itself sits beneath a monster home. Compared to other magic wildernesses of my youth, Clayoquot has done well. Yet it's less than a shadow of what it was thirty years ago. There are big plans to preserve what is now called the Clayoquot biosphere. I wish them the best.

Meanwhile the small fishing villages of Tofino and Ucluelet have become "world-class" destinations. Industrial tourism has set its sights on them.

I was born into a generation that was intellectually conscious of ecology, the same generation that caused the greatest destruction of the natural world in the history of the planet. We witnessed the triumph of the "I" over the "we." Despite our big talk we turned into a generation of looters, nearly united in a celebration of our unseemly wealth at the expense of the planet's future.

In 2005, quoted in the *Washington Post*, the American ambassador to the United Nations, John Bolton, insisted that the words "respect for nature" be removed from the United Nations charter. These three words, among others, were deemed detrimental to the world.

Our politics are not about hope any more. They're about ignoring consequences. A nation of giant people like landlocked whales are climbing into their SUVs, and they are angry — wanting more — because a world based on the accumulation of things will always mean there can never be enough to accumulate.

Three decades after that storm I keep waking up from my hard dreams, stripped naked again, tossing my muddy, soggy clothes to the dirt, my gear ruined, the rain pounding down, and the wind howling against my bare skin — heating myself with my rage and a hangover and my stupidity.

And it's becoming more and more evident, that we are all drunk on the last, unspoiled beaches of a dying planet, cheering as the fires go out.

Beloved

– ALEXANDRA MORTON –

I first laid eyes on my beloved Broughton in October 1984. It was a dark fall day, with a light but persistent rain falling. Scimitar, matriarch orca of the A12 clan, was leading her family into the area to hunt the dog salmon of the Viner River. I was by now deeply familiar with this family of whales, but the place they led me was new. We passed the Steller sea lions lounging and arguing near Duff Island, hung a right and entered Fife Sound. Fife felt narrow after the broad smooth expanse of Queen Charlotte Strait and, on a day as grey as this, the landmasses melded together, making it difficult to distinguish islands from headlands.

I laid the chart on the canopy which sheltered my young son in our Zodiac, found the sea lion rock we had just passed and pressed my finger onto the paper. As we moved I inched it forward. I was afraid that if I lost my place on this paper, I would be lost a long time among the maze of islands closing in around me. Typical of the orca, these whales had a plan. Scimitar took the southern shoreline and the others veered off towards the northern shore. Their calls echoed many times against the steep, hard submerged rock cliffs as they kept in close touch. Staying near Mum is more important than food or safety to orca society.

Brief spectacular pursuits by black whales after silver salmon interrupted the steady progress deeper into the inlets. It was exciting to see the whales in new habitat. It was exciting to be so nearly lost, to follow whales into the unknown. Droplets of moisture beaded on my handsome young husband's full beard and his eyes shone with discovery of this place. Robin Morton lived to film orcas, and I knew he found this setting irresistibly beautiful.

As raindrops — quicksilver darts of cold — slipped through the seams of my coat and streaked down between my breasts, I checked my little boy. Only three years old, he was such a good lad, so patient with his parents' eccentric need to always follow, follow, follow the whales. Through storms, over whirlpools, in diamond seas formed by the sunny westerlies, he played in his canvas tent. I had more toys than a toy store and hid them, rearranged them and made his home as fascinating as I could, but still it was a tiny space for a boy.

I first realized there were other humans in this watery labyrinth when the distant whine of an outboard motor reached my ears through the underwater microphone. We had come out the east end of Fife Sound and now entered an amphitheatre-like bowl of mountainous surrounds where new channels intersected like the hub of a bicycle wheel. The sensation was vast. I felt tiny and cold and began looking for a place for us to camp, when I spied a warm curl of wood smoke rising from a floating house. The bay was snug, a long white beach at the head, a steep wall of rock defining one edge. There a young mother stood on her dock with two shy girls peeking out from behind her legs. I felt my son's small arm curl around my knee, both timid and prodding me to approach the strangers.

I had thought children raised in the wilderness would be shy and had worried about my son, but I discovered the opposite. Children who are raised in isolation enjoy meeting peers so much that they become masters at introductions. As we guided our little boat towards the family, the eldest girl called out across the water to my son, "Hey, do you play cards?"

After a shared dinner of canned salmon, rice and homemade chapattis we slept for the first time in Echo Bay. The next morning we slipped away through wisps of fog, back to our live-aboard, *Blue Fjord*. We were home.

I am a biologist, an artist, a mother. And now they call me an "activist." Some call me much worse because I fell in love with a place slated for destruction. Business plans drawn in foreign countries decided a grand experiment would be played out here to the death of place or the venture. I can only report to the shareholders: *be aware of the woman protecting her home.* Don't expect her to give up, she does not know how. Violence is not in her toolbox, but all the attributes required to raise a child most certainly are hers — persistence, insight, alliance, inability to abandon.

The Broughton Archipelago gripped between Knight and Kingcome Inlets was once home to perhaps 10,000 First Nation people. This place generously offered exactly what they needed to thrive. Today, places like Broughton should be held sacred, since they provide clean air, water and food. The term "fresh air" here means air freshly made, radiating out from billions of leaves and needles. Rivers sluice clean oxygenated water into our reach, slaking a thirst we cannot deny. Salmon pulse out from the heart of this archipelago and back: a bloodstream of life, nourishing all that are near. If we lose too many of these sources of nourishment we lose our place on earth.

I took one sniff of this place and knew my habitat. I feel lucky to know where I belong. But nothing in this world is static; change is constant. Broughton was vast and strong when we met for the first time, but now she is maimed and convulsing. At first she held me and my children, making it possible for us to thrive; now I offer my hands, try to hold her, repay the gift. I, too, have been changed since our meeting: widow now, silver-haired with a tracker's eye, mother to a young man and a daughter. I arrived here before the makers of synthetic fish, the feedlots now anchored within Broughton and choking off her blood flow. Because I got here first I know where the pieces now lying scattered used to fit. I know these waters belong, not to multinationals, but to the orcas of "A" clan. I know that *Iwama* the humpback whale fishes Cramer Pass in March, where the people of before are buried under the sound of hurricane-force winds in cedar branches. I know the orcas should be here on the summer's ebb tide. And I believe the wild places of today will only survive if they are loved by us.

So many people stand in my shoes. Virtually every biologist who has made a life among the wilds faces the challenges that have become my life. These are fascinating times. The "average" person thinks they have no power when in fact they are the very energy source driving this runaway animal: us, the human. Every time we exchange money for things, we feed the process that made that thing, be it organic farmers or corporate superpowers. The corporations that have come to feed in my home have enormous bellies. They are so massive I cannot fully see their shape. With feet in other places, their massive mouths dip into pristine cold waters to suck up the life force here, laying waste to a perfect system.

The corporate animal will never take responsibility. While it feeds in our three dimensions, it exists only in two dimensions. The evolutionary crucible that has shaped all other life can exert no guidance on this beast. I can see its power explode the life-bearing gears of my home ecosystem like throwing a rock into the carburetor of a race car engine. Sparks, life, toxic algae and death fly out from the impact, a connection sanctioned by human law, but breaking immutable natural laws. Like touching both ends of a battery together, massive energy escapes to be sold, then nothing is left, nothing for the future, nothing for our children.

I love my species. How can I not? We are spectacular and silly like a four-year-old. But I also love my home. Broughton is pulling a veil of grey across her body as I write. Summer westerlies are losing their annual dominance to the winter southeasters. Sandhill cranes are migrating over the west entrance to Fife Sound as they have always done. I have my firewood undercover, my winter sup-

An orca surfacing with the white spray of its breath clearly visible.
(PHOTO: CULLEN PHOTOS, DREAMSTIME)

Two orcas coming to the surface for air. (PHOTO: SEARAGEN, DREAMSTIME)

ply of salmon stored away and I have allowed my best kale plants to "spawn" a carpet of food for my little daughter and me. I love the summer and the freedom it grants: a nomadic lifestyle. With my dog and daughter I have spent the last few months following whales in Blackfish Sound, but now I long for my little woodstove and the long dark nights. The southeast storms force quiet days. If we are lucky, one of the Gilford Island wolf packs will settle for the winter months within earshot so we can listen to them howl at night.

My research has shown that salmon feedlots in the Broughton, with their underwater acoustic harassment devices, forced the whales to leave. Ten thousand years of history, since the last glaciers receded, snuffed out. Many of the orca matriarchs have died since this exodus and so perhaps only I hold the knowledge of how their clans fished this place. But I cannot tell the whales. Atlantic salmon have invaded these Pacific rivers, a tough scrapper of a fish, the sole and dominant salmon in the Atlantic now loosed into Pacific systems where five native species of salmon evolved together in balance. This reminds me of the origin of the word "sabotage," the act of throwing a wooden shoe into the gears of early machines.

When I first came here I could smell the return of the pink salmon. So many leaped into summer breezes, their vital scent picked up and carried on the wind. Today I see the loss. I know tiny splashes should sparkle the length of Cramer Passage, but the water lies flat, barren as a womb without a seed. Pink salmon are one of the greatest natural inventions. These tiny salmon are the most abundant and carry massive amounts of ocean nutrients up the hillsides, where they are fed on by bears, wolves, mountain goats, toads, eagles, fertilizing everything on this coast. Because they are the shortest-lived salmon and feed low on the food chain, this fish is one of the cleanest proteins left on earth, cleaner even than other wild salmon. Pinks are designed to feed the masses: human and otherwise. But corporate business plans have decided the people can't have this fish. Instead, sea lice escaping from feedlots are eating all the young pinks. My research on the issue of fish farms is published in the top scientific journals of this planet, but deaf ears and blind eyes allow the destruction to continue, in an economy of madness.

I love my raincoast home more than I can say. Twenty-three years after first breathing in its scent of sea and forest and salmon I can only hope I serve this place as it has served me. And there are others like me. We who ventured into wilderness, made it our home, learned the secrets and the connections are not necessarily the best equipped to speak. We sought only to listen. However, Broughton is my home. Broughton was generous in her treatment of my children. I will not abandon her. I don't know how.

Echosystem

– DARCY DOBELL –

It's a beautiful June day in Tofino. I'm new to town, as yet unaware that beautiful June days are uncommon here. The sun lifts fragrant steam from the cedar planks of the deck, its warmth cut by tendrils of cool breeze that wind like seaweed around my ankles. I stand on my toes and lean awkwardly over the railing, straining to see past the corner of the house for a glimpse of the cove across the street. Somewhere out there stretches the green mosaic of forests and inlets and mountains. Behind my back, a bird calls: a single resonant weep, *like a drop of water echoing in an underground chamber, then an upward spiral of three rippling notes. I feel suddenly and strangely at home.*

When my husband and I moved to Tofino it was less for any good reason than because there was no reason not to. A bottle of wine and a late-night walk on Chesterman Beach was all the deliberation we needed to leave our fashionable urban neighbourhood.

I'd spent time in Clayoquot Sound before. I thought I knew what I was getting into.

I soon learned that the spiralling song that seemed to me such a part of this place was the call of a Swainson's thrush. This speckled forest thrush is a migrant whose breeding range encompasses most of North America. On this coast, it lays a clutch of mottled eggs among the dense thickets of salal, salmonberry and devil's club that border streams and the foreshore. The particular bird I heard calling on that first day would have arrived in my neighbourhood just a week or two before I did, and would be gone by September. It was disconcerting to think I could be made to feel uniquely at home by a nondescript bird that was really just visiting, a bird I might have encountered in the shrubbery anywhere between here and my birthplace in Massachusetts. It was also troubling to admit that, despite years spent on the west coast, I'd never heard this bird sing before. Or that I had just never noticed.

I grew up in the city. For years, I lived half a block from the ocean and never knew when the tide was high. I was aware of the weather only in the most immediate sense: it was raining or sunny, warm or cold. Here I notice the shape of clouds massing on a distant horizon, or a shift in the wind, and anticipate the approach of weather systems. I am aware of the cycle of the tide, the phases of the moon. An ephemeral stream behind the corner of my deck tells me precisely when the forest floor becomes saturated with winter rain. My body responds to these rhythms. If it doesn't rain for a few weeks I feel parched. After a dry spell, the first patter of rain on the deck comes as a relief that registers even in my sleep.

It's easier to track the moon in a place where city lights have not blotted out the night sky. In any event, we pay more attention to our surroundings when we're backed into a tight corner of land with only one road out. But there is more to it than this. When we encounter, day after day, natural forces that have yet to be buried in culverts or concealed by skyscrapers, the knowledge of where we are permeates us. It awakens a part of our being that, for a lifetime or for generations, has been waiting for the opportunity to make the link between where we are and what we do. It's the part that longs to say things like, "The wind has shifted to the west. The weather will be better for fishing tomorrow. We'll leave on the ebb tide at dawn." Whether or not we are conscious of it, our awareness of local forces stirs an internal compass that orients us to the world.

In this way, we are like migrating birds. Back in the deep recesses of thrush history, a few wandering individuals chanced to nest somewhere slightly outside the range of the rest of the population. The new location had advantages: better forage, fewer predators, or a more suitable evening temperature. The nesting was successful, the offspring wandered too, and over millennia there emerged a species of thrush that flies thousands of kilometres twice each year. The migration no longer has anything to do with a chance decision and is instead guided by such directional cues as the angle of refracted light, the Earth's magnetic field and the positions of stars. The flight is now embedded so deep in the hereditary material of the Swainson's thrush that molecular geneticists can trace the 10,000-year-old path of receding Pleistocene glaciers, and the advance of forests in their wake, by the birds' oddly circuitous migration route. The birds, presumably, are not aware of this. At some level, however, they know the position of the sun and the pull of the Earth.

My own deepening sense of place is not entirely a comfort. My growing attachment to these surroundings rubs sharply against the knowledge that I may not stay. It was more or less a given when we decided to move here that we would come and then, sooner or later, we would go. It was an understanding neatly summarized by one long-time resident who responded to my new-in-town rhapsody about the beauty of the coast and the charms of Tofino with a wry, "Yeah, well. It always makes me laugh to see the transients come and go." At the time I appreciated the irony. Now it is gut-wrenching. This place touches a part of me I did not know was there, stirs passions I never knew I possessed. I no longer want to leave. It is a puzzle: how do you make a home out of transience?

This is not an idle question. I have to balance my desire to be here with responsibilities to profession, family, community — all the practical duties that maintain the fabric of human lives. I spend a lot of time away. At times I have thought irritably that I might just as well have stayed in the city and painted a mural of Clayoquot Sound on my wall. In these moments I feel a kinship with the thrush: we are only just settled in one place when it's time to pack up and go again, and when we get there it's nothing but racing around feeding the kids until it's time to return. In the spring my own home takes on the appearance of a thrush's nest, as untended brambles of salmonberry and salal stretch over the deck. I look out from this thicket and imagine singing the same spiralling song. For the thrush, it is a breeding call: plea, advertisement and territorial claim all in one. I hear it as a reflection of our shared longing and frustration, a lament for all the scattered effort of our lives. *I desire. Look at me. This is mine. For now.*

There are times when I wish I could live in a floathouse. This is impractical for so many reasons that I can't even begin to enumerate them. Even so, I like the idea of being out of town, off the grid, buoyed by the tide, immersed in the cycles of sky and weather. I like the thought of distilling all this hustle down to a few essential rhythms. So I take the landlocked option, and walk in the forest instead. When I stop, a bee comes barrelling out of the silence. The bee zooms around me in a crazy figure eight, one loop of which encircles me, over and over again, first low and then higher and higher until it finally spirals away and is gone.

Scientists can talk about a forest in terms of its layers — floor, understorey, canopy — but these terms have nothing to do with a forest's surfaces. The dense cloud of foliage around me resolves into angled leaves and mats, shards and fronds. I once owned a sweater that my husband claimed was a shade of green not found in nature. Since moving here I've learned that there is no such thing. Beside me on a downed tree the lichens are almost white. Within just a few metres, the shadows give way to near black. Between is a complex array of greens, from soft olives to eye-jarring limes to translucent jades. Some are upright, some droop, some wave in the slightest breeze and others are stiff and glossy in the light. Hemlock boughs reach out fans of needles in soft arcs, barely moving, like fingers stroking a sleeping kitten.

Around me the forest rains. Damp air collects around each leaf and needle, then condenses into droplets that merge, trickle and fall. All through this forest the foliage combs water out of thin air. It's a reminder of what can precipitate when we simply stand still, palms open, receptive rather than grasping. A reminder of what it means to be attentive. In my old urban neighbourhood several cars sported the same bumper sticker: "If you're not outraged, you're not paying attention." Certainly, it is outrageous the way we treat the thin busy film of life on this planet, in the way we treat each other — but if we really pay attention then we have to be amazed.

This world is an astonishing place, and we mostly don't notice it. Perhaps this is necessary. Paying attention is a risky proposition. It's one thing to listen to a forest thrush sing, and quite another to think for a moment about exactly how that particular song, the colour of a feather, and the urge to wander are all encoded into a stretch of DNA and stuffed into a microscopic cell that becomes a mottled egg that releases a thrush. If we really thought about this, or if we stopped to notice the millions of little speckled birds that surge like a wave over our heads twice each year, we'd never get anything done. We would be unable to talk about anything else, all day long, or else we would be awed into perpetual gobsmacked silence. "Amazement" means literally to lose one's senses, to become bewildered. It carries the risk of what the philosopher Alfred North Whitehead calls "misplaced concreteness" — an error of mistaking the thing itself for the essence of the thing. A confusion, for example, of location for presence.

The stillness of a forest is deceptive. It would be very different if we could hear the work of plants. Scientists do not yet understand the process of transpiration — that is, how water is drawn up a plant to evaporate from its leaves or needles. The cohesion among water molecules can explain the first dozen centimetres. A combination of suction from above and root pressure from below accounts for another few metres at most. Nobody knows exactly how water reaches the top of a fifty-metre fir, but it is not a passive process. Plants actively pump water from the earth to the sky. This is not the quiet of comfort or solace, but rather a massive effort at tranquility that marks the difference between rootedness and immobility. The word transpiration itself implies a movement of spirit, somewhere between inspiration and expiration. It is a synonym for becoming.

Even the deep silence of the rainforest is less a matter of inactivity than of acoustics. Every leaf, every needle, and every droplet of water scatters sound waves. Every surface creates minute echoes that interfere with transmission. A bird song in the forest fades quickly. Unless, that is, the bird strikes the right note. The calls of forest songbirds have a characteristic acoustic pattern that differs from the calls of birds that evolved in open spaces. The song of a forest bird is specifically adapted to take advantage of the forces around it. It is a synthesis of effort: rather than silencing the song, the forest amplifies it. Every surface is a soundboard that can make a note just a little louder, make it last just a little longer.

I push through a curtain of salal and step down to the foreshore, balancing on mottled rocks that shift and roll beneath my feet. From the forest edge, a tangle of cedar and salmonberry leans toward the flooding tide. Across the inlet a mountainside shoulders its way past fingers of mist, ridges rippled like muscles under a thick green pelt. The thrush calls again and again. Instead of scattering, the song coalesces and resonates: what I desire and where I am and what I create become one exuberant spiralling call. *I am alive.*

Being Nuu-Chah-Nulth

– NADINE CROOKES (KLIIAHTAH) –

My name is Kliiahtah. I am Ahousaht, one of the Nuu-chah-nulth First Nations whose traditional territory encompasses the heart of Clayoquot Sound. I have always felt inextricably linked to the land and waters of my home. As a youngster, I remember spending solitary moments on a beach at Kelsmaht (Vargas Island) and feeling a sense of well-being and true belonging. However, I never understood that feeling as a cultural and spiritual connection to the place itself. I know now that this was a mental block stemming from my childhood anxiety of being "Indian." I was scared of being labelled "dumb," "uneducated" or "worthless" because of the colour of my skin.

Being "Indian" was not an easy thing for me as a child. Somehow I bought into the stereotypes that still permeate society — drunkenness, stupidity, even savagery. In those early years I absorbed readily the negative feelings that were projected towards me, choosing to believe that I was not as deserving as others. My adolescent years passed with no strong sense of identity. I was not totally lost, though; I had a group of dear girlfriends who loved me as I was. Their caring and devotion helped strengthen my self-esteem.

I was also fortunate to come from a very loving and supportive extended family. My grandparents especially had a significant influence upon my values and ultimately upon the person I am today. The role of grandparents in Nuu-chah-nulth culture is integral to the sustenance of our people — grandparents are often responsible for the rearing and teaching of children. When I complained loudly to my grandmother that I was bored, she would always answer, "It is only you who makes yourself bored." This answer used to frustrate me, but inevitably I would pick up a stick and start drawing pictures in the sand. Now I understand that she was trying to teach me to be self-reliant, in order to create my own happiness.

Today, when I am feeling sorry for myself and cursing my lot in life, all I have to do is remember some of my grandmother's experiences. After a couple of moments my own challenges are quickly put into perspective. Although she told me many stories of turbulent and trying times, one always stands out for me. My grandfather was of Irish descent from the east coast of the United States, and on a road trip, after a long drive heading home towards Ahousat, Grandma and Grandpa decided to stop and rest overnight at a motel in Port Alberni. Grandpa went into the office to register for a room and the attendant happily obliged. As Grandpa was filling out the form, the assistant happened to glance towards the car and spotted a brown woman sitting in the passenger seat. He promptly grabbed back the registration slip and informed Grandpa that the motel was full. The funny thing about when Grandma told this story was that she had no malice in her voice. It was told matter-of-factly, in a way that taught me to refrain from taking on other peoples' issues. It was a matter of personal dignity. Grandma taught me much — we liked to stay up late and talk about everything. Some of my most cherished memories are the many weekends spent at my grandparents' house.

Intertidal life, Frank Island, Tofino. (PHOTO: JEN PUKONEN)

Cow parsnip (*Heracleum lanatum*). (PHOTO: JEN PUKONEN)

Grandma remains one of my greatest role models, teaching me to find the good in every situation. I once asked her about her experiences at residential school and, as I was bracing for a heartbreaking story, she began relating anecdotes with humour and goodwill. Despite the fact that she was removed from her parents for a great deal of her childhood and adolescence, she managed to find the positive aspects of the experience. Of course, she was one of the lucky ones who did not suffer abuse or beatings. Her attitude of recognizing that we cannot change the past leaves her free not to dwell upon it. Instead, she demonstrates a greatness of spirit, an enduring optimism that transcends life's most difficult experiences.

It was not until I became a young adult that I began to connect consciously to my culture. This connecting has been more than a life-enriching experience; it has been a life-altering one. One of the greatest gifts my elders gave me was a sense of belonging, purpose and self-worth. There is no particular memory that defines the turning point — the reconnecting. Rather, there is a series of recollections and, more importantly, feelings towards people. I had the opportunity to spend two glorious years working on a traditional-knowledge research project, during which I interviewed numerous Nuu-chah-nulth chiefs and elders. These people made me feel accepted, loved and comforted. They helped me understand my role in the community, and they gave me vocabulary to unspoken and deep-seated feelings that I had always had. Thus I received a new understanding of what has always been within me.

Those intimate and loving experiences have since defined who I am today. I now cherish a connection to both place and people. Crucially, I have been filled with an overwhelming sense of hope for the future — for our children, our people and the role we as Nuu-chah-nulth must take in the modern era.

I am proud of who I am. I know that it is not how others define me that matters, but how I define myself. These teachings are the underpinnings of Nuu-chah-nulth belief. The Nuu-chah-nulth culture is founded on valuing the entire community, where each person fulfills his or her own role contributing to the greater good.

Our way of life is also premised on the concept of a "human family"; that is, my actions are integral to the interests of my family, my neighbours, my community and the Earth that we depend upon. My culture teaches me to be responsible for my actions: they are my choice, and therefore the consequences are also mine to own.

A respected elder and friend once shared a story with me of his adolescence. In his youthful exuberance he went out duck hunting one day. He returned proudly to the community, his canoe loaded with ducks. But he had taken more than he needed; in fact, he had taken more than his entire community needed. Instead of the praise he had expected, he was reprimanded for his serious lack of respect and taken by the elders into the woods for solitude, prayer, redemption and re-teaching. After a few weeks, he had relearned the meaning of *iisaak* (respect), including the errors of arrogance and greed. He was then reintroduced into our community, reformed. At great expense, his family hosted a potlatch acknowledging his wrongs and celebrating his deliverance. Although he made his amends, his error was recorded for all time in a song that has been passed down from generation to generation — a song of accountability. He may never know how profoundly his story has impacted me personally or helped me to understand the importance of humility and respect.

When I am in the forest and close my eyes, I can feel my ancestors' presence. And there I find peace. There is no word in Nuu-chah-nulth for "wild" or "wilderness." There is only home. Elders have taught me to respect the lands, waters and people within our traditional territories (which often follow natural watershed boundaries). These territories are essential to our well-being. Our beliefs are thousands of years old, yet perhaps more relevant today than ever: *hishuk ish ts'awalk* refers to interconnectedness. People are dependent on the environment and upon each other. All is interconnected. Deeply rooted in this perspective is personal accountability for the decisions we make. It puts the onus on us to make choices with humility and respect. This is what being Nuu-chah-nulth means to me.

Finding Home

– JOANNA STREETLY –

It is the hormonal duty of teenagers to spread their wings and escape from home, so it is ironic that when it was my turn to take off, I arrived. I left England behind and arrived on Vancouver Island: Campbell River in the rains of January 1987. No Silk Cut cigarettes for sale anywhere; boyfriend left behind; tense long-distance phone calls in the presence of friendly-but-strange people who cleaned their kitchens obsessively and put away the kettle when they were done with it.

It wasn't long, however, before I moved out to Strathcona Park Lodge, to a tiny room beneath the eaves of a large chalet, overlooking Buttle Lake. As it was the off-season, there were only a few people living at the lodge, but it was fun because we ate and worked together and I actually made a few friends. In between working and eating, I crouched on my balcony feeling quite Canadian, smoking Players Light cigarettes, watching the clouds gather and move between the mountains and the lake. I felt at ease because I was used to isolation, having grown up surrounded by sugar cane fields in Trinidad. Although the quietness and the liquid beauty of the lake fascinated me, it wasn't until I joined four others on a two-week-long kayak trip in Kyuquot Sound that I really found home.

We paddled from beach to beach, seldom seeing other human beings, occasionally locked on tiny islands by the storms. I felt intensely connected to the wild coastline and the water that smoothed it all together. It reminded me of the north coast of Trinidad: steep rainforested slopes diving down to black rock shorelines and remote sandy coves. All that was missing was the heat and the howler monkeys. It was as if the almost-forgotten geography of my early years had crept back into my heart. And as the ocean moved beneath me, and the feel of it travelled through my paddle and into my bloodstream, I felt that I could dive into this new world and never return.

In camp, my crouching post became a driftwood log where I rolled my own Drum Mild cigarettes (Dan, the guide, smoked Drum Mild) and stared at the water with intensity — seeing everything and nothing all at once. It was the end of June and my return to England was approaching fast. The dim cage of university hovered at the edge of my thoughts. The boyfriend weighed heavily, like a sea anchor. I tried to keep everything in the present tense — to avoid the future at all costs. The future involved decisions, decisions I wasn't sure I would be strong enough to make.

After bouncing back down the logging roads at the end of the trip, we stopped in White River for lunch. We ate at the diner and it was there — over hot dogs and beer and the realization that we all smelled really bad — that I vowed to spend as much of my life outdoors as possible. The feeling washed over me in heavy waves — a teenage epiphany. My future decisions suddenly became easier, guided by the principle that I should never, under any circumstance, work in a tower block or office building, away from daylight and air; that to do so would be a betrayal of the highest order — *a betrayal of my soul!*

It's hard to believe that almost twenty years have passed since that summer of self-discovery, but here I am, on a floathouse in Clayoquot Sound. It's further south than Kyuquot Sound, but it's still on the west coast of Vancouver Island and I'm still paddling a kayak as often as I can. My crouching post is now a smoke-free window seat, but I feel most alive when I'm paddling along the outer coast, swell crashing on the rocks around me. I still thrill to the smell of wet salt air and the moments when clouds shimmer and ripple on glassy water.

These days, I travel over water and through life with a little girl by my side, seeing her delight as she catches the wind with her chubby little hand, or ploughs through sand on all fours, heading for the tidal pools. I hope that a vital connection to this place is surging through her year-old body as she does these things and wonder if fate will cause her to lose the feeling, the way I did, and have to rediscover it later.

These days, when I say the word "home," I am referring to this floathouse. It's where I keep the physical trappings of life: a bed, food, water, a cook stove, a woodstove, books. It's where my partner, Marcel, and I find each other at the end of the day and where our daughter sleeps with the terrible trust children have that the world will still be safe and beautiful tomorrow. This floathouse is the enclosure to which I long to return, a place where I know peace will settle on me as I walk through the door.

It was Polly Adler in 1954 who said, in a book by the same name, that a house is not a home. But when I consider my body's reaction to place, it seems to me that a home is *anywhere* you have discovered or rediscovered love. Eight years ago, when I first moved my floathouse to this remote spot, I was grieving. A long relationship had ended. I was alone, far away from other people. I went to work in town during the day, floating along in my altered world, but by day's end, the tension in my ribcage and upper back would send me home rigid. Not until I walked through the door of the floathouse would my body relax, my lungs open, my shoulders sag.

Back then, I wrote poem after poem, scorned food and rolled Drum Mild cigarettes. I crouched on the doorsill for hours watching eagles, ebbing tides and rain through the smoke I exhaled. I woke early and sat in bed for hours, drinking tea and staring out of the window at the trees on a nearby island. Slowly, out from the heavy weight of my crumbled relationship, *I* emerged, unfurling like a fiddlehead in spring. At this floathouse I was able to recognize and appreciate all the different types of love I had been given; I was able to let this love come in and go out freely, the way it should. Love began to exist, for me, as the collective pool that it is, not some well-outlined *thing* that I should channel to a single place or person.

In the same way that love should not exist within boundaries, so the definition of home should not be confined to one exact spot. My frenzied teenage passion for the coast, and my desire to make it my home, makes sense when I see it as a rediscovery of my father's love of water and wild places. Few people could be so excited by a bird or butterfly, a day in a boat: fishing, snorkelling or swimming. "Look! Look!" he would whisper, his eyes as bright as the feathers of the jacamar he had just spotted, perched on a branch. "Shhh!" he would gesture if he had just found a zebra butterfly pressed flat and camouflaged against the bark of a pomerac tree. I remember his hand beckoning through water when he had spotted a particularly beautiful angelfish, or a scary moray eel. He was just as fascinated by dangerous creatures and could produce exciting shark stories like coloured handkerchiefs from a pocket. When particularly moved, he would sigh with delight and declare something to be "a lovely, lovely, lovely thing."

Of course, the flip side of such enthusiasm was a terrible temper, and Dad could stomp up and down and cuss with the best of them. But I knew to stay away from him when he was like that, so I mostly remember the happy times. While Dad could find loveliness anywhere he went, Mum was busy cultivating it at home. Wherever they lived, Mum quickly started planting trees, digging ponds, taking cuttings, shaping her surroundings. The length of time they were planning to live in a place never stopped her. Even if it was a temporary home, she didn't want to lose an opportunity to create beauty. She worked with fervour, so it was never long before we were eating fruit from trees she had planted when we moved in. My mother is an artist. She sees things through the unfocused squint that she learned at the Royal Academy and she often stops to notice beauty in haphazard, unexpected places. Her appreciation of these things is less theatrical, but just as deep as Dad's was.

Even though a love for nature is something I believe to be nascent within everyone, I imagine that, through observing both parents, my own appreciation of nature was enhanced. It must have grown strongly in me while I was a young child, until everything changed. There was a period of my life which I call the Dark Time. It began when I was eleven and I was sent away to boarding school in England, while my parents' lives were in flux. They were moving between Trinidad, Tobago and England and I seldom returned to a place I knew as home. My grandmother insisted I spend vacations at her big old house in the English countryside, but although she wanted me there, she did not love me, and I was just passing the time until I went back to school. And school . . . was school. Rules and walls and closed rooms. I was socially inept — had difficulty making friends. The cool, damp climate kept me plagued with bronchitis and asthma. I floated through school life, semi-engaged. I learned sarcasm as all teenagers do. I argued with my mother. I started skipping lessons. I took up smoking. I took up drinking. Eventually, to my mother's dismay, I found a bad-influence boyfriend. My world grew narrower and narrower.

I shudder when I imagine the directions my life could have taken from that point. So many doors I could have walked through. I don't think I was aware that anything was missing, but in retrospect I can see that something was: I was homeless. I could not find home within school walls, nor could I find it arguing with my mother about a boyfriend whom we both knew wasn't good for me or the career choice that I was unable to make. When I was seventeen, the only thing my parents and I agreed on was that I should take a year off after school and travel. We all thought I would be escaping from home. None of us saw that I was searching for a home. None of us dreamed that I would find it so quickly.

Now I look around me at my small, simple house. I look out of the window at the family of eight otters playing on the dock. I remember standing in frosty winter moonlight watching a plump little harbour seal, three feet beneath me. It was draped over the breakwater log, sticking its heavily whiskered nose in and out of the water, blowing bubbles and very rude raspberries. I remember many meetings — glances exchanged — with wolves, and I remember the summer day when I swam past a bear that was swimming the other way. My mind scrolls through countless wildlife events. Wildlife aside, I think of the hours I've logged simply gazing at the water, words and ideas eddying through my brain. Living here has helped me to form a unique, powerful bond with my natural surroundings.

Last month, my mother said to me, "You live such a wonderful life. What will you do if you have to move?" I answered quickly, flippantly, because her question took me by surprise. It was the first time she has ever acknowledged her regard for my home, and I was taken aback. My mother has lived an exciting life, full of wild places, adventures, love, loss and beauty. She moved far away from her family to do this, but has always bemoaned the distance between herself and me. Before this moment, she had always diminished my floathouse by calling it a boat and lamented my dull, untravelled life and the absence of cultural events. This time, however, there was a wistful note to her voice. I could feel her recalling all the adventures she had with my father when he was alive, wishing the clock back. And here I was, boating with my daughter, having epic moments, living life the way she might have.

Later, when I had time, I thought about what she had said. I've always known that my lifestyle can't last forever. I don't plan things for the long term, but when I try to imagine living somewhere else, I draw a blank. It would be like amputating a limb, and I would always feel the loss. I've tried to live in town before, and I've always come away feeling useless and empty. Even one night in town is frustrating enough to send me fleeing home in the boat, like a junkie seeking a fix.

My mother's question followed an event that took place earlier this summer, which also caused me to contemplate the loss of my home. Late one evening, after Marcel and I had finished supper, washed the dishes and were enjoying the summer twilight, my cell phone beeped a message alert. As usual, the noise was intrusive — an obnoxious antidote to the calm of the plate-glass, light-filled water. I sighed and checked my messages. And there was Dave's quiet voice telling us that there had been an earthquake in California and a tsunami warning had been issued.

In recent memory, there have been many warnings on this coast, but few devastating tsunamis. I took a deep breath. I could only think of one thing: my daughter, lying asleep in her bed upstairs, eyes closed, arms flung wide. I pictured a giant wave sloshing through the floathouse, taking her away from me, from life. A cold stone grew in my chest. I tried to fling the vision from my eyes, but it lingered and I saw how deeply and how quickly my mother-love has become embedded in this home. I picked up the first aid kit and a plastic jug of water, preparing to take them to the boat. Marcel took out a Rubbermaid tote for us to fill with essentials. We turned on the VHF radio and listened to the Coast Guard weather channel, all the while discussing our plans. Just then, the warning was cancelled due to lack of evidence. The earthquake had not spawned a tsunami. This time.

Whenever I think about losing this precious home, I have to remind myself that it can happen at any moment (an earthquake one morning as I wash the dishes; a terrible winter storm that blows the house to bits; a notice, evicting my anchors from government foreshore). I remind myself that my passion for nature will help me to recreate a home if I can find a suitable place. And even though I dread the loss of these walls that I proudly nailed together years ago, I have to be prepared. I have to live each day as if it's my last. And so I do, by skipping thoughts of the future and doubling my appreciation of the water and the wildlife. But in a silly, upside-down way, this strengthens my love for my home and its surroundings. My attempts to "let go" make the idea of "letting go" even less palatable. It's like trying to prepare for the death of a loved one: you can tense your body, tighten your heart and speak banal words about letting go and moving on, but you can never be truly prepared. You can only deal with change when it happens.

When I think about being homeless, I am reminded of Antoine de Saint Exupery's *The Little Prince,* in which the little prince falls to earth in the Sahara desert and cannot get back to his planet. He mourns for his friend, a flower, who may die without his care. "If you love a flower that lives on a star, it is sweet to look at the sky at night." He misses the sunset, which he can see at any time on his little planet, just by moving his chair: "One day . . . I saw the sunset forty-four times!" The ultimate test of his love for his home comes when he arranges for a poisonous snake to bite him so that he can go home without his body, which is too heavy. This is not something he does easily. He is frightened, but he carries through with it.

In this exquisitely sad and beautiful story, the prince's body is his physical home. He describes it as a shell. But away from its setting — his planet — his home becomes unliveable. In a less dramatic way, life at my floathouse would be so much less meaningful if there was no wild setting. Even though my house *is* my home, it is nature, all along, that has fostered my ability to find love and happiness here. That makes wilderness and nature the greater home force. I need to be able to trust that, as long as there are wild places on Earth, I will be able to make a home and be fulfilled. After all, I began as a teenager, acting on a dubious cocktail of impetuousness and longing. It is only in the last few years I have begun to see the path behind me. And to help me find the path ahead, I can take advice from *The Little Prince.*

"Here is my secret," the little prince tells us, "a very simple secret. It is only with the heart that one can see rightly; what is essential is invisible to the eye."

Uncharacteristically Simple

– CHANDRA WONG –

Down a twisting, turning, heart-stopping highway lies the remote village of Tofino, a destination of choice for people locally and from around the world, a place in which to be surrounded by the wilderness of ocean and temperate rainforest. Myriad colours soothe the eye from the turquoise and aqua of the ocean to the jade and emerald of the spruce, hemlock and cedar trees. Chartreuse moss glows beneath the forest canopy, where pale green fiddleheads unfurl their new growth in a lush and moist understorey. Whether the precipitation is a fine mist or heavy downpour, jewel-like drops hang suspended from branches, with fungi and lichen sparkling in the dim forest light.

A million people visit Tofino every year, yet finding reasonably priced housing can be nearly impossible on this peninsula, where land is at a premium. Film, sports and music stars own beachfront mansions which stand empty most of the year; few homes sell for under a quarter of a million dollars and rent is equally outrageous. First Nations reserves are crowded and frequently troubled with unclean water; the Tla-o-qui-aht people are actively seeking expansion and a secure future in their homeland.

Not surprisingly, then, there is something about the people of Clayoquot Sound. They are not your typical, run-of-the-mill folk, but ever willing to step outside the box and face each challenge with ingenuity and determination. Without luck or wealth, sheer stubbornness can help in the pursuit of finding and keeping a place called home — even if it's in a ditch.

As the cliché goes, necessity has mothered inventive solutions, despite the objections of town fathers and mothers who cite violations of sanitation, fire and building codes. One controversial route is to follow a trail into the forest and set up a tent, squeeze into a hollow tree, or even lie down in a forgotten culvert beside an old logging road. I have stumbled across a surprising number of squatter sites in the bush. Unfortunately, everything in a rainforest quickly becomes damp and clammy. What hasn't started to decay with mould and mildew may soon be covered by slimy mud. And campers have been known to leave what amounts to truckloads of garbage behind them. The by-law officers are inevitably kept busy throughout the warmer months of the year.

These limited, potentially uncomfortable, illegal and even dangerous land-locked options may explain why many have looked to sea for a solution, in boats and floathouses. Living on the water means a close connection with the moods of that lady they call the Pacific. The gentle lullaby of wavelets at the edge of your home may rock you to sleep with oystercatchers and sea lions as your roommates; or, you can awake with the shrieks of gales in a fierce winter storm, leaving you ragged, exhausted, maybe even white-knuckled after enduring what felt like a never-ending night of being tossed and turned.

A couple of my friends lived on a converted trawler for a time, the *Skookum Queen,* a small vessel that shifted with every movement of the sea. There was barely enough space to turn around. When one of them was cooking, the other could only sit at the table or go outside. When the bed was unfolded, they had just enough space to lie flat, shoulder to shoulder, and that is no small feat for two people both over six feet tall. Somehow, they managed to squeeze their lives into this tiny boat. Later they were able to opt for a life on land, buying and renovating a Winnebago, which they christened "the Toaster."

Two other friends lived in a fishing camp in the nearby community of Ucluelet. The large wooden building was built on pilings over the ocean, and here fishing boats transferred their catch, already packed in ice, to transport trucks waiting to take the catch elsewhere for processing. When I first heard about their new home, I imagined a huge warehouse dwarfing their couch, TV and kitchen table, where everything was permeated by a slightly fishy odour. But they occupied what had been the manager's cosy apartment. Sea lions could be heard barking in the harbour as the ocean broke against the pilings. Communal areas, like showers and laundry facilities to wash away the aromas of work, and a sauna to chase away the bone-chilling damp of the sea, were once heavily used by workers, but now served only my friends. They had a housewarming party, creating a dance floor in the unused ice room. Mini-Christmas lights were strung from the beams and along the drain channels in the floor, while a turntable played disco music and a whirling disco mirror ball hung from the ceiling. Instant "Fishothèque."

The draw of the "Wet Coast" is strong. Compared to the challenge many others face when trying to find a place to live in Clayoquot, my own experience was uncharacteristically simple. I merely interrupted a conversation next to me in a café where I was having lunch.

"Sounds like you have a place to rent?" I boldly asked.

"Yeah, are you looking?" responded a man in a charming Québécois accent.

"Yes."

"Maybe you should come by."

My new place was warmed by a wood stove. I had only a bed, table and chairs, but it was home, my home.

I was one of the lucky ones.

EDITORS' NOTE: *Like other remote destination hotspots, the Long Beach peninsula suffers from scarce affordable housing. The Tofino Housing Corporation hopes to address this problem by taking over forested lands to build multiple types of "attainable" homes. Workers who serve the area's thousands of visitors need a secure and healthy base. The more a place is loved, sometimes the more it is altered.*

Called Home

- HELEN CLAY -

I have been on a quest to find my home. I thought, as a child, that home was a single specific place. I thought that it was in England. I thought that it had walls. Now I'm living in Tofino, on B.C.'s wild wet coast. I am blissfully happy and completely amazed to be here, despite all the effort it took. Now I know that home has walls made of tree trunks and falling rain. Home is the place where I feel a deep sense of belonging. Home is where I am welcomed, valued and received with love.

I travelled around the world on a mission to find home. Home, I felt, would be somewhere rich with natural landscapes, somewhere I felt at ease. On a visit to stunningly beautiful New Zealand, I gave serious thought to making *it* my home, but something did not seem right. Something inside, a small voice I've learned to listen to, was saying "no, not yet, there's something else you need to do first."

FIRST TIME VISIT

A very dear friend of mine suggested I should take a trip to Vancouver Island, and visit a town called Tofino. I hit Tough City on a December afternoon.

It started simply. After I stepped off the bus, I wandered round the town. An hour later the bus went past again, and the driver waved. This may not sound like a big deal, but where I grew up on Exmoor, everyone knew everyone else and their cars too. So you got waved at a lot. People stopped their cars in the middle of the street or lanes to have a natter, broken up only by the arrival of a tractor. If you changed your car, you felt very lonely for a week until people realized it was you. Until that Tofino bus wave, I hadn't realized how much I'd missed being waved at.

That night, there was a talk about the world temperate rainforest network by Pat Rasmussen. I learned a lot about the rainforest, but what impressed me most were the people I met. December is not exactly peak tourist month, so the hall was full of locals of various vintages, all of whom were deeply passionate about Clayoquot Sound.

Here were people who shook hands firmly, smiled with sincerity, looked you straight in the eye and said "Pleased to meet you" with feeling. Everyone was strikingly individual, yet they came together to make a clan of outstanding people, here because they wanted to be. I met several amazingly talented writers, artists, photographers, musicians, bird experts, and an ecopile of environmentalists (if that is indeed the collective noun). Many people had moved here from other parts of Canada or further afield, so I understood I wasn't the only newbie in town.

The next day I wandered through Tonquin Park to the beach, sat on a rock and listened to the Pacific rolling in. Those waves have a very special way of breaking: a long, slow, inevitably mesmerizing sound. I hiked the Meares Island big tree trail the day after, with a birder; we spent two and a half lovely hours gently exploring. I loved the old hollow stumps. And the hanging garden tree. I laughed as I turned my face upward to the canopy and was roundly dripped on by serious rain. The broad green salal

Blue-eyed-grass — a fitting name because the blue "eyes" appear from the side of a grass-like stem. (PHOTO: JEN PUKONEN)

Banana slug — the second largest species of slug that is native to the temperate rainforest of the west coast, the "vacuum cleaner of the forest."
(PHOTO: JEN PUKONEN)

leaves shone and the nets of lichens held raindrops like diamonds. The sound of the drips falling was enchanting. It felt like the right place to be.

I walked on North Chesterman beach too, both on a quiet evening when the glory of the clouds reflected on the damp sand, and on a stormy day when the crashing of the waves took my breath away. All alone at the Botanical Gardens, I was taken by the sculptures, the forest, the awesome Jan Janzen driftwood buildings.

Here in Clayoquot the air moves differently. The breezes really caress you. The tide can sidle in so gently and surreptitiously you hardly notice it moving until your feet are wet and your backpack soggy. On a clear day, the scenery is simply breathtaking. On a day of rain, horizons close in and a cozy feeling envelops the town . . . a good time to sit in Breakers café with a hot chocolate and watch the world go by.

In the middle of a city I felt more acutely alone than I ever do in the middle of the rainforest. For me, this is because in the forest I can be completely and utterly myself. There is a great stillness in the quiet places. There I am aware of the undercurrent of joy that bubbles up within me, irrepressible. Out in the "real world" I am sometimes accused of being naive — and if I spend enough time out there, watching the news and hearing about all the corrupt humans, I feel pushed to the point of admitting that there is no point to existence. But then I move out into nature. I focus on a spider or a slug that has no knowledge of George Bush, and I feel renewed. More than that, I feel this will endure, that human life is ephemeral, untested, untried, not working out too well . . . and I know I'm only as naive as the planet, which keeps putting out glories day after day despite what we do to it. Under the category of glories, I like to include slugs and raindrops (every single one unique) as well as more obvious wonders like whales. I cannot seem to stem my love for this Earth, or accept that humans as a species have no choice but to trash it.

The most interesting part of my response to Clayoquot lay in how I felt about the people. I had always thought I was a bit of a loner, that beautiful surroundings would be all I needed. New

Zealand was certainly beautiful, but it did not call to me. In Tofino I instantly got along with people and felt of value to them. I felt welcome, wanted, acceptable and accepted. I had not expected to feel this way. I came here looking for beaches, storms and rainforest, and I wasn't disappointed — but I hadn't realized how captivated I would be by the humans belonging to the place.

The week after my visit, back in Vancouver, I found what had previously seemed a nice small city was now a busy rushing metropolis. I found I was beaming widely at people on the street — and receiving blank stares in response. Robson Street gave me an anxiety attack. Too many shops, people, fashion. It all seemed shallow. Thinking back to Clayoquot, I realized I was suffering all the symptoms of falling in love. I couldn't eat or sleep. I was disgustingly happy. My family thought I'd joined a cult.

I knew my path was leading to Clayoquot. I could feel it guiding my steps. Events I would previously have written off as coincidence now seemed to be steering me. One day I was aimlessly browsing in a second-hand bookstore on Richards street, a great old-fashioned store with higgledy-piggledy piles of books everywhere. My hand came to rest on Betty Krawczyk's book *Clayoquot: The Sound of My Heart* (1996). I got the message. I was going. At the top of my "to do" list was "Check I'm not dreaming." I wasn't. Unfortunately, this meant backtracking to England: my house must be sold, a visa sorted, and my ass returned to Canada as fast as possible.

COMING BACK

I said my goodbyes to Exmoor by treating myself to a few days in a little cottage deep in the woods, near the sea. There I walked the South West Coast Path along cliffs that dropped vertically to the sea below. The endless blue of the sea merged with the endless blue of the sky, and I felt as though I could step off the narrow path and fly, on and on into the blue. On the tops of the moors I lay on warm springy heather and grass nibbled short by the sheep. The sweet coconut scent of the gorse and the bees buzzing gently lulled me to sleep. I knew that if this new adventure didn't work out, Exmoor would still be there. I could always come back. But I also knew that Tofino was something I had to try. I could not live out my life wondering "what if?"

It was different when I returned: summer, the town dustily overwhelmed by thousands of tourists. I lived at the Botanical Gardens, where I worked three days a week. I also worked for the Friends of Clayoquot Sound. Two jobs I loved and believed in. David Pitt-Brooke, in his beautiful book *Chasing Clayoquot* (2004), says "Environmental activism is like weeding a garden: the job is never done." I knew exactly what he meant.

August, known locally as Fogust, broke up the sunshine and I loved it. I woke each morning to the dripping leaves, a quite different sound to rainfall, as the trees gather the moisture from the air and gently deposit it upon the waiting forest floor. The fog burned off early each afternoon until the evening fogs returned. On the beach one day, my head was in sunshine but my feet were enveloped by gorgeous long tendrils of drifting fog, rolling in from the sea.

I wanted to try kayaking, and began my paddling training with a three-day trip to Vargas Island. To start with, I was heavy-footed on the rudder. So, while everyone else was paddling in what appeared to be a straight line, I would be wandering in a big lazy S-shaped path: oops too close, bear right. Oh, too far away, bear left. Oops, too close. . . . And so on, all the way past Stubb's Island until it came time for our fast push across the busy lanes of Father Charles Channel. I seemed to be a much faster learner at dinner that night, where the task at hand was Dungeness crabs. I got real good, real quick. This seems to be a constant in my life when it comes to food.

We hiked across the island to Ahous Bay, munching on berries along the way. We picked whole stems laden with salal berries and ate as we tramped. There were red huckleberries and a few early evergreen huckleberries. The guide taught us about Labrador tea, sweet laurel (good) and swamp laurel (bad). The interior of Vargas Island has a bog area, rare in Clayoquot Sound, where the trees don't grow very tall. There's a lot of shore pine, unusual inland. The king gentians were just opening, a beautiful blue.

Club moss made a sprawling claim along nurse logs. Birding was a little thin: only some chestnut-backed chickadees and a talkative Steller's jay.

Ahous was a huge sweep of sand, and we were the only people on it. As we fanned out, I walked along the shore, watching the waves so full of sea lettuce they broke green. There were three common murres paddling along, looking around at the waves, then dipping their heads under to search for dinner. An osprey caught a fish, only to be swiftly pursued by an eagle that harried the osprey until the prey was dropped. I could not help but weep at the beauty of this place: that I should have made it here, freed from a life of loneliness amongst many to find companionship amongst such breathtaking isolation. And then I looked around at the empty beach, the Pacific Ocean stretching away, and started laughing uproariously, all by myself, simply unable to believe my luck.

NEXT STEPS

My position here is tenuous at present. A year's work permit ticking away. Residency takes a couple of years. Will Clayoquot Sound want me to stay? I hope so. You see, I know there is a secret in the deep woods. I know there is a mystery — that magic dwells here. I know it when I stand on Meares Island, in the old growth forest, and hear the stillness. I feel it as the furious pace of my little existence slows down to match the beat of life in the great, old, silent trees. In here I find the consolation that is given by the stars: that not one moment of our brief, hectic lives matters; all the frenetic effort we put into fitting in and making money and acting such-and-such a way troubles the aeons not one bit. Nonetheless, I still believe that we humans do have a role here on Earth; as long as we respect our home and cherish it, we are fulfilling our part. We are the Universe, looking at itself.

Yearning

Excerpts from

The Last Voyage of the Loch Ryan

- ANDREW STRUTHERS -

I sat up late that night in the wheelhouse of my fishboat, long after the rest of town had gone to sleep. I had a serious decision to make.

But my mind kept drifting. There was something hypnotic about the pool of light that lay on the chart table. The warm breeze blowing steadily through the porthole smelled of low tide and piling tar. The smell of home. The red eye on the shortwave radio blinked, and a tiny voice flitted into the wheelhouse like a blackfly: "Auntie, are you on this one?"

It was a little girl over in Opitsaht, the Native village across the channel from Tofino, which some local wag had dubbed "Opposite."

Reprinted by permission of New Star Books.

Silence.

The radio's eye winked again, and her auntie came on: "Yes."

Now how did Auntie know she was the right auntie? That's the mystery of a small town. Each human life is invisibly connected, like roots tangled in the duff beneath the ancient rainforest, a forest so vast and thronged with creatures that humans find themselves busted down from tyrant to citizen. But don't let insignificance bother you. Out here, even the constellations get lost among the lesser stars. Preconceptions break down and rust in the back of your mind. Redskin, redneck — those are just words. What is real are these ghost voices on the radio, a little girl finds her mother's sister in the ether, while water slap-slap-slaps against my hull

silence

the smell of cedar. The scent of diesel. In the morning, white sheets of mist lie tangled between giant trees, and some old guy standing on the dock says, "It'll burn off by noon." Which it never does. Stunted spruce combed sideways by the sea wind. A yellow eye-stab of lichen on grey rocks. Welcome to Clayoquot Sound. Fifteen years in the arms of this goddess, and tonight is my last.

The Plan: leave Tofino before dawn. Ride five miles out on the falling sea. Catch the tide as it floods back into Juan de Fuca Strait. Make Victoria by midnight. Drop anchor across from the Legislature. Read a list of demands through a bullhorn. They're going to be sorry. You can kick a man out of the Clayoquot, but you can't kick the Clayoquot out of a man.

Some are born to adventure, some find adventure, and some are dragged ass-backward through adventure, screaming for mercy to gods they no longer believe in. I'm in the last group. Still, it's better than a day job.

After university I tried living "on the grid," with a day job and a credit history and a NOTICE OF DISCONNECTION and a stress-related skin disease, but it just wasn't me. So I moved into the forest in Clayoquot Sound and built a pyramid out of cedar and glass, perched on a hundred-foot cliff, looking out through a canopy of giant trees over a sparkling limb of the Pacific.

On the horizon lay an island where the ancient village of Echachist once stood, until it was destroyed in battle two centuries back. For years there was no sign of human activity down there. Then one evening the setting sun caught on two golden cedar beams. Someone was raising a house frame.

I asked around the Common Loaf Bakery. The builder was Joe Martin, whose forebear had been the chief at Echachist. I watched Joe's house go up, and just before he finished the roof the Federal Government slapped a demolition order on my door. My house wasn't up to code. I moved into town for the winter, and by spring the fuss had died down, so I carried my stuff back up the hill and continued my contraband lifestyle.

To pay the bills I worked on fishboats in summer and at the fish plant in winter. My one attempt at a career was writing, but it didn't work out. I loved the writing but hated the career. These days you can't simply write, you also have to be a celebrity, which I find unsettling, because the only celebrity I resemble is Shrek.

"Hey, Art! The *Loch Ryan* sold!" said Peter.

"Finally! Well, just tell whoever bought it there's no live-aboards. We already got too many, we're getting complaints. Who bought her?"

I stuck my head out of the fish hold. "I did, Art."

"Oh. Morning, Andrew. Well, you can stay till you get her fixed up."

Art figured I was okay because I got up with the sun, a habit even deadbeats like me develop when there's no electricity. Art had the habit bad because he had been born miles off the grid at Hot Springs Cove, where his dad owned the land the springs are on. It seemed like Art had been the wharfinger in Tough City since the world was young. He ran the dock out of a shed at the top of the ramp, where the Fishermen's Club met for breakfast at six every morning. The menu never varied: ten cups of bad coffee and a million-dollar view.

The only person who was supposed to be living on the dock was Shorty, a pint-sized Russian with pierced nipples and a cowboy hat, who ran a knife shop out of an old fishboat moored on the second finger. Each knife was a piece of art honed from hand-forged steel with a scrimshawed bone handle. His boat was stuffed full of tools, paintings, carvings and giant bottles of Demerara rum. Art had grandfathered him into the paperwork because the Feds allowed one live-aboard per harbour, to cut down on vandalism.

But scattered between the first and fifth fingers of the dock were at least a dozen others. Next to Shorty lived old Johnny Madokoro, whose dad had taught the settlers how to troll for salmon back in the thirties. Johnny had a house up in Port Alberni and lived on his boat while he was fishing. When he was a young man he had owned property over on Stubbs Island, but he was of Japanese descent, so when the war came the Feds seized his house and tore it down. I said, "I know how you feel."

Lashed to a log off the end of the dock was a double-ender with a cabin on top. In it lived Lance, his wife and their baby. Lance had grown up in Ontario, where his dad took to beating his mom. One day the old man started up right at the dinner table, so Lance knocked him out cold with a ketchup bottle. His mom said, "You better be gone when he wakes up." Gone he was. He made his way west until the forest came down into the sea, and now he jigged cod and drank all day.

On the fourth finger, a mandolin player called Ryan lived in a big green sailboat with no mast. He was a chef by day, and every night a ring of hippies sat on his cabin roof plucking bluegrass.

At the very end of the fifth finger, moored between Peter and the *Loch Ryan*, was a giant wooden seiner called the *Oldfield*. She was painted black from bow to stern except for the windows and looked like she could take out an icebreaker. She started life as a halibut boat up in Prince Rupert in the thirties, then became the tow-off boat for herring season in the Clayoquot. Now she was a pirate ship. The hull seemed to be formed from hundreds of coats of lumpy paint, the crew subsisted on a steady diet of gooseneck barnacles and beer, and every night there was a party on board with reggae pumping from the wheelhouse at nosebleed volumes.

I began to wonder if I really wanted to go back to university in the fall. There was too much fun to be had right here in the Clayoquot, firing guns in Barry's basement and learning to operate giant pieces of machinery. The beauty of the place was, there were no rules. You could do anything your mind could conceive. So I built a sixteen-foot scale model of the Great Pyramid in front of my house on Chesterman Beach. While I was working, the new cop puttered up on a motorbike. It seemed like there was always a new cop in town. I guess the Feds didn't want them getting too close to the locals. He said, "You're not going to live in there, are you?"

I said, "It's solid sand."

He said, "That's okay then. It's just, there's too many people camped on the beach. We're getting complaints."

❧

[EDS. One of Andrew's adventures is bushwhacking with other local characters in search of a rumoured standing stone on the side of a mountain. The real homecoming turns out to be the view: one of lives, familial and cultural roots re-emerging from the landscape.]

A hundred feet above the cedar grove the ground broke into mossy chunks. I followed Doctor John along a ridge, gripping the springy limbs of dwarf spruce that might be a century old. Up ahead we heard a cry. Ten minutes later we found Adrian and Jimmy sitting with their backs against the base of a gigantic column of rock.

John pulled out a huge tape measure and started measuring.

So analytical. The column was formed from volcanic basalt, which must have cooled slowly, crystallizing into a pillar. The cliffs at the top of Lone Cone on Meares Island looked the same. What made the Monolith unique was that the surrounding basalt cooled rapidly, forming weaker rock that crumbled away into soil and left the column to jut like a stone thumb.

Two blocks of basalt lay tumbled against the Monolith's back like giant dice. Adrian scrambled on top of them to get a group shot and found a narrow ledge that slanted farther up the sheer face. Soon we were all edging along it. Halfway up there was a gap in the ledge. I could see promising handholds on the far side, but Adrian turned back. His steeplejack days were over. So did Doctor John. A shattered shinbone up here would be both irresponsible and inconvenient. Jimmy glanced at the drop, paused, then lunged across the gap and disappeared round the side. I followed.

Around the corner the sun was dazzling. Jimmy hauled himself up out of sight. The top was flat as a table. I sat, feeling the lichens tickle my calves, gazing out across Millar Channel to the Atleo, round to Hot Springs Cove and down the coast to Raphael Point. I could see half my life from up there. At the Sulphur Pass blockade I'm chased naked through the forest by the new cop. In the hot springs I sit, lashed by rain, warm and happy on a freezing February night. Five miles out to sea I shift from foot to foot at the wheel of my fishboat. In a derelict cabin on one of the outer beaches I curl against River, listening to the endless rain.

Jimmy could see his whole history etched deeper in the same landscape. At the mouth of the Atleo, Copper Woman cries so hard that Snot Boy comes out of her nose. At Raphael Point, Jimmy's great-granddad harpoons the last sea otter. Battle cries rise from the palisade at Opnit. In the blue distance beyond Hesquiat, the *mamaalthi* arrive in floating houses. It was an epic moment. And not a cloud in the universe.

I said, "This place is amazing. How come there's no stories about it?"

Jimmy said, "Maybe there are stories. How'd you think we found it so easy?"

Trail to Hot Springs Cove, 1987. (PHOTO: JULIE COCHRANE)

Floating home covered in cedar shingles at Hot Springs Cove, 1987.
(PHOTO: JULIE COCHRANE)

[EDS. Another adventure — an act of mischief — hearkens back to a less developed Tofino, a more innocent time when wild animals could still raise their young beside the house where Andrew's ex-wife and daughter lived.]

"COME ON DOWN AND ENJOY THE VIEW!" said the sign outside the pub. One night, on our way home from a play rehearsal, Mike and I changed ENJOY to DESTROY. We laughed and laughed. Then a light clicked on in the pub office, so we ran and hid in the patch of rainforest behind Barry's place.

The forest was magic. The creek that ran into the sea beside Gwen and Pasheabel's house had once fed a dozen giant cedars, most of which had been bucked into chunks the size of tool sheds and tumbled into the gully. The forest canopy had grown back over top, and now the gully was pocketed with caves made out of wood. A cougar made a den in there one year and raised come cubs. Pasheabel loved to explore the grotto with a flashlight.

The creek ran down from Barr Mountain on the other side of Campbell Street. For years I thought only one of the giant cedars still stood, the one that towered above Gwen's house. One morning Barry and I sat on his deck watching eagles flap around the topmast branch, shrieking like evil squeeze toys as they are wont to do when making love. Barry said, "I love that noise. So long as there's eagles in that tree I figure the town will be okay."

Facing the Mountain

- CHRISTINE LOWTHER -

An infant harbour seal cries for its mother on the rocks flanking my neighbour's oyster farm. I pause to listen more deeply to the night, the cold water enflaming my skin. Then, scissoring my arms and legs, I make angels with the ocean's bioluminescence, comet-tails streaming between my fingers. The Milky Way faded when my glasses fogged up in the hot tub, but this is an even better light show. Floating on my back, I know what Mum meant in her poem "Song" about feeling the galaxy on our cheeks and foreheads.

Later, as I fall into bed, the seal is silent, and barred owls call to each other in the forest. My bed is level with the open window; as I face the mountain, my hip and shoulder mimic the curved contours of the land. A stanza from another of her poems comes to me:

And nights when
clouds foam on a beach
of clear night sky,
those high slopes creak
in companionable sleep

The last sound I hear before sleep, and the first one upon waking next morning, is of fish nipping at insects, repeatedly breaking the water's surface for a fraction of a second.

I live in a floathouse: a cabin that floats upon the sea. Under the summer sun, I immerse myself in water, alternating between hot tub and cold ocean. If I am kept from this ritual, spending too much time in town at my job, extreme grumpiness ensues. The streets are choked with tourists, friendly, relaxed people whom I've been serving all day. It is in town where I am marooned; it is out here, seven kilometres up an inlet, where I relax.

Out. Here.

Only by entering the ocean and feeling complete immersion do I become part of this bay, forest and mountain. For years I was curious about my obsession simply because not everyone shares it. Many are happy to soak in the hot water without ever putting so much as a toe into the cold. Then Chief Earl Maquinna George, in *Living on the Edge: Nuu-Chah-Nulth History from an Ahousaht Chief's Perspective*, described how for his people (whose territory is just north of here), immersion in wild, cold waters was "cleansing of the soul, cleansing of the body, cleansing of the mind, cleansing of evil spirits . . . exchanged for the strong will to live and fight."

For many of us, it is a fight to make Clayoquot Sound our home. Other than squatting, floating is arguably the only affordable way to live in the district of Tofino. If a new government floathouse tax comes into effect, even that option will no longer be open.

An hour and a half in a scratched old kayak or twenty minutes in a barnacle-infested motorboat: these are my commuting choices. In a storm, the motorboat's the only choice, and those twenty minutes can be the longest of one's life. Once I arrive, there is nowhere else to go, not even for a stroll. Long ropes anchor the floathouse to the densely forested shore across the water, where black bears sometimes meander, munching on shore grass, salal berries and huckleberries. They'll take a dip only rarely, and they are not interested in humans.

The natives used to dig for clams here at low tide, but non-native oysters colonize the mud now. The oyster farm lies across the mouth of the bay.

The floathouse, which I named Gratitude, is tucked into a far corner of what my neighbour calls "God's Pocket." It is my refuge, my sanctuary, and the place where I come face to face with myself. How could one not feel grateful here? The creek never dries up, providing drinking water and powering a micro-hydro generator acquired from a friend who runs an alternative energy business. The summer sun heats a large solar panel, or "collector," that warms the homemade tub next to the greenhouse, and the greenhouse alternates as an informal sauna. Golden nugget tomatoes grow well, as do basil, cucumbers, celery, leeks, and strawberries. This is sane abundance. Tofino's abundance leans in another direction: uncurbed growth measured in metres of parking and square feet of mansions and condos. There is less habitat each year for wildlife, and people leave their garbage out; bears are attracted, become a threat to people and are "destroyed."

The inlet's contrasting quiescence seeps into one's bones, soothing. Friends visit from their cities, enter Gratitude's parallel reality, and find themselves sleeping better than they have for years. A harbour seal and a river otter sometimes share the space under the deck; eagles and kingfishers catch fish outside the window. A great blue heron lands on the dock, glares with its deadly, perfect round eyes, and deposits its white splat-signature. Sy Montgomery spent quality time with emus in Australia. "Staring into that intelligent, shining brown eye," she confessed, "I felt as if I were capable of looking directly into the sun."

With water, however, comes movement and unrest. The horror of sinking lives in the back of my mind. After one November storm, my position turned ninety degrees overnight; two kayaks were up the creek, two solar panels were in the chuck, and the micro-hydro generator had been dragged and damaged.

In summer, the horse flies can be particularly bad. Yet I find the hardest thing to bear is the sense of invasion when sailboats or yachts anchor a stone's throw away in a Sound full of equally beautiful locations. I'm here to escape from civilization. When tourists paddle their dinghy around my cabin, staring through the windows, my own anger disturbs me.

Yet I am also an interloper, the offspring of colonialists. Meares Island is Tla-o-qui-aht traditional territory. It encircles Gratitude, snugly embracing it, and also forms the dominant viewscape from Tofino. The island is unusual because it has never been clear-cut, apart from a patch on 730-metre Lone Cone. After aboriginals and other local people turned loggers away in 1984, the Tla-o-qui-aht First Nation declared Meares a tribal park. It remains tied up in court with the lengthy treaty process. The Sound is a UN–designated Biosphere Reserve, yet the entire place remains unprotected from any industry.

For this precarious paradise my love grows deeper every year. Dragonflies mate in the air, spiralling down like a doomed airplane. Bees bend back their hind legs to groom their wings. Moon jellies (*Aurelia aurita*), during their annual reunion, join together in one long mass between the house and the dock; I can run my hand across the gelatinous, slippery surface. While I am swimming it is impossible to avoid colliding with dozens of them, but they don't sting. It rains more than three metres annually here, but sometimes when I think rain has come, dimpling the water from shore all the way to the farm, it's the moon jellies pulsating, tasting the air on a fragment of exposed flesh.

❧

After growing up in Vancouver, then spending five years in England, I found that my activist leanings led me here, to the west coast of Vancouver Island. I was led to something I'd never heard of: temperate rainforest. Here were immense western red cedars as old as 1500 years, possibly even 2000 years, upon which the moss appeared to grow as thick as the walls of some castles I'd seen. My fellow Canadians were clear-cutting the trees for a New Zealand corporation.

Hiking past waterfalls on the Walbran River, I gaped at creamy Salvador Dali-esque canyon walls reflected in jade-green water. Beyond a campsite dubbed Giggling Spruce, a landlocked lake supported its own exclusive species of trout. Some of the cedars were natural works of art themselves. With their gargantuan candelabra crowns, they looked like multi-headed elephants, every trunk raised. I recalled my mother's profound love of the arboreal realm:

> Trees are
> in their roots and branches,
> their intricacies,
> what we are
> ambassadors between the land
> and high air
> setting a breathing shape
> against the sky
> as you and I do
>
> . . .
> Trees moving against the air
> diagram what is
> most alive in us . . .

As scientists searched for endangered marbled murrelets and activists gave tree-climbing workshops, I began to feel at home in the rainforest. Addictions to junk food and certain television shows fell away. Poems about roots and bugs seemed to write themselves. An unmistakable sensation of time slowing down permeated my blood, and a week went by without use of a mirror. Being a temporary inhabitant of an ecosystem that took ten thousand years to evolve stretched and calmed my mind. Yet the "curly opaque

Pacific/forest, chilling you full awake/with wet branch-slaps" was how Mum had described B.C.'s mountain woods in her poem "Coast Range." Indeed, the slow time of the coastal temperate rainforest kept me awake and alert to both the beauty and the wrongs of the world.

In the summer of 1992, I was arrested with about sixty others for blockading logging activity in Clayoquot Sound. The decision to do so came naturally; my mother had once led me in protesting the destruction of two aged and giant trees near our city home (to make way for development). In 1993, more than 900 arrests were made in Clayoquot. It was the largest act of civil disobedience in Canadian history.

I was sentenced to house arrest. This was many years before the floathouse. I was still transient — living in a tent. Suddenly I had to secure a home immediately. It was time to settle down at last — a thought which made me squirm. The island's west coast is wild, arguably romantic, but never rural, never quaint like an English patchwork countryside. Was this where I was supposed to be? Settling down could only imply facing, rather than packing, my baggage.

It wasn't easy to become a permanent fixture in a town rebuilt for the wealthy. Moreover, while there were bitter feelings between Tofino and the logging town of Ucluelet to the south, Tofino suffered polarization within its own boundaries. All this because of trees. Innocent, oblivious trees were ultra-political. I had made enemies before spending a single night in my new home town. Mum would have entirely understood.

When I said "Tree"
my skin grew rough as bark.
I almost remember how all the leaves
rushed shouting shimmering
out of my veins.

At first, my partner and I shared a house with several roommates for six months, knowing that we would have to move in the spring when the bed and breakfast season resumed. Then, miraculously, we were offered a log cabin to ourselves, five minutes from town, on the corner of a forty-acre island covered in rainforest. Out the back door stood giant old-growth cedars, hemlock and sitka spruce. In front was an unheard-of view: a multi-peaked mountain draped in uncut original rainforest. Mount Colnett's flanks swept down to the prime bird habitat of Lemmens Inlet's mudflats. Such an inspiring environment made it as easy as it was necessary to meditate on the front deck each morning, before heading into town to work in a busy whale watch booking office.

At home my partner — whom I'd met on a blockade — hooked up my laptop to a solar panel. In this way I was able to write when I was not chopping wood, heating rain water, or composting our waste. After five years on the little island, the cabin was sold and we were evicted. Tofino had become even worse for average-income tenants, so my partner bought the floathouse — something I had never heard of before coming here — for $7,000. It was a real fixer-upper, too. There was no escaping this place. On the contrary, I couldn't spend enough time here. I was always needed at my salesclerk job in town. Every moment floating became precious.

Before we even began to unpack, I put up the hummingbird feeders. An immature rufous hummingbird joined a feisty troupe of females, its beak much shorter than the adults'. Mature hummers zoomed up and surprised us; the juvenile took more time, approaching in a sort of aerial waddle, backside swaying under tentative wings. Sometimes it would perch on the flower it was feeding from and forget to fold its still-outstretched wings. Nevertheless, this young bird knew what it had to do and was determined to practise and succeed. It was also not as concerned with its surroundings as the adults were, and would come quite close to us in pursuit of sustenance.

The young hummingbird often perched on the clothesline, wobbly, plump and downy, looking at the flowers, ignoring us, opening its beak to peep like a nestling. Was its mother among the adults? The others abandoned their combative conduct at the feeder to give it a wide berth, where it perched at one of the four plastic yellow flowers. The most obnoxious bird (named "Rammer" by me) stationed itself on the line and drove everyone else off. Oblivious to the noisy clashes all around, the baby buried its bill in the plastic flower, its entire body rhythmically guzzling sugar water. When Rammer finally lost patience with Baby, the little one retreated to the line and peeped pitifully, its bill opened extraordinarily wide. My partner was eating a slice of watermelon, and Baby wobbled down to try a lick of that. It tried a nostril, too, with near disastrous results.

One autumn day last year a family of five river otters was feasting on perch, flicking their tails as they dived, sending up bubbles, and popping up with the shiny, wriggling fish. This went on for an hour, leading me to believe the bay must contain thousands of perch. When two smaller otters approached the float, I took up position bent over a gap in the planks on deck. Sure enough they were swimming close to the surface, around and past my flotation logs, just under the gap. One young whiskered face emerged between the water and me, and I said softly, "Hi." It disappeared and returned with its sibling; the two of them looked up at me for several seconds. After that, there was a great deal of conversation between family members out of sight, as well as the usual din of crunching and lip-smacking — river otters have no table manners whatsoever. They were obviously curious animals, and one emerged beside the dock, looking cautiously around for me. I froze behind a large planter of lemon balm. Not seeing me, the otter dived. All was quiet, and I retired indoors.

About a quarter of an hour later, I glanced up from my windowside table. The whole family was dog-paddling steadily toward the house, five heads held high above the water, a live crab in every mouth. At six feet from the dock they dived. Under the greenhouse deck, they feasted with audible gusto.

As I sat on the deck one spring afternoon, a lone river otter bubbled up from the deep and drew breath. Swallowing some morsel, she clambered onto the dock, shook herself and commenced rolling on the warm, dry wood. Then she jumped to her feet and chewed her bum like a dog with fleas. Either she didn't know I was sitting beside and slightly above her, or she didn't care as long as I remained still. I barely breathed. She used one foot to scratch her head, while the other pointed straight up, like a cat's. I could see the webbing between each clawed toe. Stretching her neck, she sniffed one of my potted tulips with her huge nostrils until a sigh of wind in the chimes startled her. Soundlessly, she slipped into the ocean — her second skin.

In "Voyageurs" in *The Nature of Nature,* Scott Russell Sanders watches river otters playing and finds he wants things from them. "I wanted their company. I desired their instruction, as if, by watching them, I might learn to belong somewhere as they so thoroughly belonged here. I yearned to slip out of my skin and into theirs. . . ." He wants from the otters the same thing he desires from everyone — friends, strangers, neighbours, his grown daughter:

> I wanted their blessing. I wanted to dwell alongside them with understanding and grace. I wanted them to acknowledge my presence and go about their lives as though I were kin to them, no matter how much I might differ from them outwardly.
>
> (Sanders, 210)

It proved to be a good week for otters. I kept waking up at four A.M. hearing splashes around the rocky shoreline. The fourth night of this, I propped myself up on an elbow and looked blindly out into the darkness, longing for some sort of visitation. Bioluminescence surrounded an animal swimming swiftly toward the floathouse. As it dived, a thin coating, a submarine blanket of light covered the body and I could see her — an otter — darting underwater here and there, a pale ghost. She finally scooted under the dock, leaving an explosion of white-green marine fireworks in her wake.

Wild creatures can seem like miracles, even aliens in a world fast disappearing under pavement. Like the screech owl that collided with my window and flapped hastily away, leaving its mouse dinner behind. Or the frog that swam, unfolding itself under water into some long, thin creature I'd never known, and climbed out onto my foot. A swimming snake, a fluttering sphinx moth with extended proboscis, a shimmering school of herring or circling, tightly striped mackerel can send me into raptures. As Natalie Angier describes the feeling in "Natural Disasters": every time she sees a creature flash across her path unexpectedly, she finds it "more exhilarating than catching a glimpse of Al Pacino or Mikhail Baryshnikov on the streets of Manhattan." For her, it's "a momentary state of grace." For Sy Montgomery, even when the emus she was studying became very familiar, she did not take them for granted or become bored. Rather, she fell in love with them. As she says:

> It was raining harder, with a cold, bone-chilling wind. My coat and hair were drenched. I didn't care. I wanted only to find them. . . . I realized that it wasn't the data I was after. I just wanted to be with them.
>
> (Montgomery, 42)

Our own desires are clear. What of their effects? Like Sy Montgomery we may want to be in the company of wildlife, but when does whale watching become whale chasing or bear watching become bear scaring? Wild animals neither want nor need my company. I have no wish to become the invader. At low tide I marvel at beings which, as far as I know, have no idea that I exist: sun stars, bat stars, pipe fish (a cousin to the sea horse), slender kelp crabs and graceful dendronotids. All of these creatures thrive in the vulnerable marine eelgrass ecosystem. The best that I can hope for is undisturbed, undetected observation of them along with infrequent benevolent encounters, even though I yearn for a much deeper connection, even communion.

Only with humility is it possible to wake up to this place and its power. Every window open, I stretch out in the hammock on a warm afternoon, lullabied by varied thrushes and Swainson's thrushes. I have always found the spiralling song of the shy Swainson's thrush the most haunting of all bird songs. In my semi-conscious state, their utterings are the voice of the forest. They come from all sides, a constant calling and answering, ricocheting around the bay. The level whistle of the varieds offers a sweet accompanying undertone.

Instantly all is silent. From the southeast shore a lone, sad howl issues forth into the still air. And again. No more.

The thrushes tentatively renew their calls, their spirals dizzying me toward slumber. I fancy there are bears huffing and rummaging among the rocks. I dream that my eyes are too heavy to open, so I can only listen. Gradually the voices of the birds grow strange, slow down, deepen, birth an echo. Though I cannot see them through my paralyzed eyelids, I feel sure the bears are standing up on two legs and their attention is on the floathouse — on me. Goosebumps spread over my skin, like a school of fish suddenly brushing the water's surface as an eagle glides over. The beings on shore are people, or spirits. But they have no message for me. On the contrary, they are questioning me.

I wake up quickly then. There is only one thrush calling now, from a distance. I am overwhelmed by the dream, which is giving rise to new questions. What is this place? The entire island is special enough to be a tribal park, so it would make sense if some of its bays, inlets, or creeks were considered sacred to the Tla-o-qui-aht people. What about this particular bay in which I live? Am I trespassing in a holy place? Something is creeping in, seizing my insides, rare and vital. This is not only Carr's "burning green in every leaf": every leaf, needle and cone has eyes. The air itself hums with alertness. Here is a glimpse into a patient, evolved dimension, more real and crucial than great cities. If the creek were ever to dry up, the sound of its silence would end the world. The forest watches warily, for Lone Cone must always continue talking "with the casual/tongues of water/rising in trees," as my mother put it.

Though I have not yet had visions, I have heard things. My neighbour Mike, working on his oysters in a gale, watched helplessly as my greenhouse floated past him at the same time that a Tasmanian Devilish waterspout spun toward him. The spout veered off at the last second, and as Mike gasped, he inhaled a small rainbow. He says he's been happy ever since. The future is unknowable; we can only breathe in the wonder that is this place now. My mother funnelled it into poems: "Look, we're riding past Venus." She acknowledged her own desire to be

> aware of the spaces
> between stars, to breathe
> continuously the sources of sky,
> a veined sail moving,

while she celebrated that

> The land is what's left
> after the failure
> of every kind of metaphor.

Growing up, I blundered my way out of despair by seizing words — my mother's, my own and many others' — and by pondering the ineffable stars. So far I haven't succeeded in growing armour on my freckled hide and, anyway, even the toughest bark can be scorched by fire. I revel in the electric chill of the astringent ocean on my skin. At home in Gratitude, sensitivity is a gift, the preferred state of mind. In the meantime, there is much work to do and it's important to find strength. Logging of old growth continues, moving into the last pristine valleys. We toiled to save a single tree in Tofino and were forced to wrap it in steel to make it "safe" from vacation condominiums constructed beneath its tilt. The world craves sanity. I am grateful for sanctuary. Let me lie down with the mountain so we can face each other's gladness, my heart as full as this miraculous planet.

REFERENCES

Angier, N. (1994). "Natural Disasters." In William H. Shore's *The Nature of Nature,* New York: Harcourt Brace & Company.

Carr, E. (1946). *Growing Pains: An Autobiography.* Oxford University Press.

George, E. M. (2003). *Living on the Edge: Nuu-Chah-Nulth History from an Ahousaht Chief's Perspective.* Winlaw, B.C.: Sono Nis Press.

Lowther, P. (1996). "Random Interview." In *Time Capsule.* Vancouver: Polestar.

———. (1977). "Song." In *A Stone Diary.* Toronto: Oxford University Press.

———. (1996). "Riding Past." In *Time Capsule.* Vancouver: Polestar.

———. (1977). "Coast Range." In *A Stone Diary.* Toronto: Oxford University Press.

———. (1974). "'At the last judgment we shall all be trees.'" In *Milk Stone.* Ottawa: Borealis Press.

———. (1968). "On Reading a Poem Written in Adolescence." In *This Difficult Flowring.* Vancouver: Very Stone House.

Montgomery, S. (1994). "The Emus." In William H. Shore's *The Nature of Nature,* New York: Harcourt Brace & Company.

Sanders, S. (1994). "Voyageurs." In William H. Shore's *The Nature of Nature,* New York: Harcourt Brace & Company.

All in a Day's Dream

- MICHAEL SCOTT CURNES -

An old tree has died in our urban back yard this year. I cannot brace it. I cannot blockade outsiders from harming it. I cannot send young activists to live in its branches. No media release will change its fate. It is dead.

This is symbolic of my current reality living outside Clayoquot Sound. In Clayoquot, you could save a tree. Hell, you could pretty much preserve an entire forest because braces and guywires and blockades and tree climbers and media releases work *inside* the eco-bubble that is Clayoquot. Outside, I'm afraid, the world seems so far gone it just doesn't give much more than a collective sigh about anything.

I don't know why this tree has died. A year ago, I pulled off the English ivy that clung to its twenty-foot-long trunk like a turtleneck. Besides *dead,* I don't even know what kind of tree it is. I don't know how long it has stood here. I don't know who planted it. I don't know if I could have done anything to change its fate. I don't know if any other person has even noticed or cares that it is dead, even though it is visible from just about every vantage point in the neighbourhood. I do know I haven't done anything about it, and that's just not like me.

Lying in my fancy, mail-order hammock, beneath the withering arms of this city-choked giant, I realize that I am parched. Rain-water does not reach my roots for the simple reason that my roots are not here. Since leaving Clayoquot Sound, I have wandered the desert that is the rest of the world. I don't feel as intensely or as compassionately about things as I once did. I don't hold out the hope for the environment and for humanity that I sincerely embraced for seven years while living in the Sound.

Here, the sun hurts my eyes and burns my skin. I am assaulted by media and entertainment overload. I consume, consume, consume. I eat fish laden with heavy metals and wash it down with an acidic punch from the rain and spring snowmelt captured and stored within the city's "pristine" water basin. There is no tide here but the flow of traffic which is constant and never recedes. I miss Clayoquot. On particularly rainy days, suspending reality and deploying Herculean imagination, I've tried to replicate my Clayoquot sense in this place but the rain here does not arrive sideways and it wimps out after a sprint of maybe twenty minutes, tops. The sun, if it disappears, is never gone for very long.

In my back yard, I stretch my body out in the hammock beneath the dead tree. I wait for the hammock to stop its rocking. I focus on my breathing and wait to achieve whatever stillness is possible in the middle of a city of more than a million people encompassed by six-lane freeways. The leafless, flowerless, needle-less, lifeless branches of the tree overhead have been imprinted by the sun onto my eyelids to cast gnarled silhouettes on my consciousness that fester like a new tattoo before the scab. I could wait for them to fade or heal over, but in my obsession with this particular dead tree, I know they will not. I have been obsessed with other trees before this one.

When I doze off, whether in my hammock or in my bed at night,

I often dream I am flying over and through and around Clayoquot. Several times every month I escape the restraints of the city and take to flight in my sleep, sometimes piloting an invisible plane like that of Wonder Woman, but most often under my own volition using a variation of the breast stroke through the sky. This particular afternoon, my dream does not waste any time in getting underway. I am rising from the hammock. As I assume my flight position, I bump the back of my head against one of the upper branches of the dead tree and adjust my stroke to avoid the others. I am keenly aware of power lines at the moment and my heart accelerates. Next to power lines, the jumbo jets taking off and landing at the major airport not far from here frighten me the most. But I am soon in the clear.

I want to stay beneath the scattered cloud cover so as not to become directionally disoriented, but I am on my way now and turning back is not as easy as it might seem in my dream state. A pudgy soft cloud inhales me into its centre. I dread it when this happens, for it is when I am most vulnerable and need to concentrate most on my stroke technique. In the blinding whiteness of the cloud, I begin to hear what sounds almost like a mournful honking, far off and muted at first, but coming closer. I soon recognize a chorus of Canada geese. I think for a moment that I might have flown with this squadron before. I vary my stroke and elevate my chin to nudge into their formation. I'm afraid I'm not very graceful at this manoeuvre and I earn squawks in reprimand as a few geese are momentarily displaced.

"I hope you guys are headed for Clayoquot Sound," I tell the goose to my left, perhaps with too much enthusiasm for a newcomer. "Clayoquot Sound is where I would like to go today," I confide to the gander on my right when I don't get a response. This flapper narrows a black eye, assessing whether or not I'm up to the trip. I sense scepticism and pull to the front of the V formation to lead the others just as we leave the clouds. Below, I instantly recognize Kennedy Lake and realize I am qualified to lead the flock at least for this stretch. This is familiar airspace.

With my body in the lead displacing and compressing a greater volume of air, I reduce our elevation in search of a tail wind. "I don't suppose the group is ready for a breather?" I ask as nonchalantly as I can. "I know this perfect spot on the mudflats ahead just past Grice Bay." It's no use, though: I wear my non-aerodynamic clumsiness on both sleeves. Why am I wearing sleeves? I am not dressed for flying.

"HONK!" Comes from my right side as the gander moves to the front spot in the V formation, taking us into a climb I can't possibly manage. No matter. I'm home now. I drop to a glide a few metres above the glistening mud and spy a pair of herons in ankle-deep water. Their stance is certainly more my pace after the long flight from the city. I land on a rock nearby and strike up a conversation.

"I used to work not far from here," I volley their way. One heron turns its steely eye on me, while the other remains statue still. "I tried to take this shortcut home one day," I continue. "You see, I thought I'd save time by walking along this mudflat shoreline from the visitor centre to Chesterman Beach."

Still, the herons ignore me. Affecting a laugh to get their attention, I reveal the error of my judgment. "Of course I can tell you now that the shoreline here takes the shape of a giant hand. My so-called shortcut had me walking all the different fingers of land — it took hours! When I finally stumbled out of the woods onto the bike path at Jensen's Bay, covered in mud, exhausted and dehydrated, I was just about as close to panic as I've ever been."

Sensing the herons aren't buying my story, I quickly stroke my arms into the air above the rock and motion them to follow. "Here, I'll show you the hand then. You'll see. Come on!"

To humour me, they lift their lithe bodies out of the water and join me in flight a cautious distance behind. I slow my strokes. Now who's humouring whom, I wonder. I smuggle a grin to one side of my face, but when I look back a few breast strokes later, the herons have ditched me. I fly solo over the giant land hand. It must be a view well known to them.

I arch my feet for a little more lift to clear a few ancient cedars that have jutted into my flyway. With a deafening *whoosh* I am joined by a swarm of fifty greater yellow legs in search of their next buffet.

I stroke faster to keep up, and we land in unison on the mudflats about twenty metres from a bed and breakfast. I know this place. I managed it for years with my ex-boyfriend-turned-best-friend. I think of calling out to see if my old friend is poking around the giant house, but I am already sinking up to my knees in this nutrient-packed tidal goop. “Hey guys,” I alert my yellow-legged comrades, “I might be in a bit of a spot here.” I have a chuckle at my own expense, realizing that these 250-gram birds are of no help to me now unless every worm, grub and mini-crustacean between here and Meares Island were to vanish suddenly in some macro-invertebrate rapture. I sludge and clop in the direction of the shoreline that runs along the botanical gardens. The mud grabs my legs with the grip of a hundred labourers' hands but I manage to reach the spindle of a fallen hemlock tree and use it as a bridge to terra-more-firma.

Brushing the earth glue from my legs, I hear the clucked musings of a raven overhead, teetering back and forth on a snag. I am somehow able to interpret his language and comprehend that I am to follow him across the channel to Morpheus Island. There is something he wants to show me, and I trust him. I try an experimental stroke to see if I can get off the ground with the added weight of the muck. It's my dream, so of course I can. The raven leaps from the branch to lead the way. Within a quarter of a kilometre the mudflats give way to an incoming tide. Just as we fly over a substantial channel of water, the raven suddenly pulls up and flaps in place, something he knows I am unable to do. I can only stroke forward, and in my momentary confusion and indecision, I stall and then begin to fall. I can hear the raven laughing raucously as I try all my aviation tricks. I arch my feet this way and that. I frantically try to stroke with my legs together and then apart. It's no use. I am falling from midair. Just then, a bald eagle swoops down in a graceful glide. I can see how her wings are positioned and manage to imitate her. My dive evens into a glide with not even a metre left between the water's surface and me. I set my chin and pull up slightly as I follow the eagle back toward shore.

We are flying along the highway toward the village centre with

Barr Mountain on our left, Mt. Colnett on the right and Clayoquot, Vargas, and Felice Islands beyond. I am rubbernecking now, which slows me down, but everything is so familiar I can't resist. If the eagle is losing patience with me, she doesn't show it as she constantly looks back to make sure I'm keeping up. Then, with a short glide and legs reaching forward, she lands on a bare branch I instantly recognize as part of the crown emerging from the top of the Eik Street Cedar Tree. I alight next to her and heave a gasp, not because I am winded by the short flight, but because I am rendered dumb by the view from this perch. I once spent months looking up at this tree from the ground. I joined the true visionaries of the town and defied development and convention and ignorance to keep this tree standing. I have written about this tree in media releases, prose and rhyme. I have photographed and painted this tree. I have sung about this tree. More than anything, I have celebrated this tree. I don't know but perhaps history might even record my hand in saving it. This is Clayoquot after all. You can save trees here.

I have to remind myself that this is a brief afternoon nap and not a coma and that I have to keep moving along if I am to complete the roundtrip before waking. There is so much I want to see. After thanking the eagle for having the sensitivity to bring me back to this tree, I hastily soar out over the Fourth Street Dock. I take up with a couple of gulls who are heading across Templar Channel to Wickaninnish Island, or Wick as it is known to locals, on a rumour that a dead basking shark has just washed up there.

For reasons that only make sense in a dream state, I am least confident about my aerial breast-stoke when flying over water. Certainly the two kilometre-wide channel between Tofino and Wick Island has given me much cause to worry in this life, what with its legendary zipper that is said to present a standing wave without warning. I've never seen this personally, even though I've made dozens of crossings by kayak and motorboat in all weather, including a thick fog one late summer evening. Still, the sea has always frightened me, mostly because I think I saw the *Poseidon Adventure* at too young an age, but I also took a tumble in a sea kayak out by

Wilf Rocks once. The former boyfriend and I went on to build a house on Wick that would forever require boat access of one kind or another. I didn't make a crossing without white knuckles around a paddle or clasping the side of our aluminum boat my boyfriend had named *Sea Leopard,* or hugging a load of milled cedar siding and beams in either Marcel's or Mike's skiffs. I could not, in the end, become comfortable as a seaman or a boater of any kind. Only my sentimentality and my dream-induced invincibility has me braced to make the crossing now.

Mid-channel, the damn gulls seem to have forgotten where they are headed or why they'd set out in the first place and begin veering left toward Echachist. Never mind a rumoured shark carcass, I am already distracted by a shiny object in the distance that turns out to be the sun reflecting off the metal roof of the house we built. We spent several years petitioning the other island residents for a lottery long shot at becoming their neighbours. Then we spent more than a year clearing salal and fallen trees from a rocky promontory fifteen metres above the sea. We'd been through so much, perhaps too much, before the day finally arrived when we engaged a helicopter to spend a day flying across the building supplies waiting at Huebner's sawmill. The house went up during a single tourist season while we managed a local resort, promoted local businesses, "inn-sat" a bed and breakfast and cranked out another edition of a local tourist guidebook. All just to pay for the privilege of building a home in Clayoquot Sound.

I can't bring myself to land on the deck that my ex built even though it means crossing the channel a second time without a rest. He has a new boyfriend who is finishing the kitchen and flooring of the house that's taken three boyfriends to complete. I am very proud to have been part of that trilogy. I arch my left foot as I circle the promontory and head back across the channel. Yes, very proud.

The swath of treeless land ahead that beckons me like a landing strip carved out of a Columbia drug jungle is a place I call "The Last Resort," where I gave five years of blood, sweat and yes, tears — lots of them — in exchange for enough writing material to last my lifetime. As I fly over, I am reminded that I cut my Tofino teeth defending the duck pond below from collar-less, leash-less, law-less dogs and the occasional mink. In the awkward house sitting on the beachfront, I consumed more alcohol than a human liver should be able to process. I wrote novels, a stage play, a musical, a dozen biting letters to the local paper and a couple of seasons worth of features for *The Sound* magazine. The rains of Clayoquot were very good to this writer.

But I need to be starting the journey back from sleep, so it's time to hitch a ride on the draft of at least a hundred western sandpipers as they compress into a MacKenzie Beach fly-by, before hopping the headland for a landing on Chesterman Beach just short of Frank Island. I continue to a spot in front of a bed and breakfast where I spent the last two years of my Clayoquot life. I was married right here to a man named Richard in one of the first fully fledged, legally sanctioned weddings of two men to take place anywhere on Vancouver Island. Out of something like 2,500 truly remarkable days that I lived and loved and learned in this place, my wedding day on this beach was my very best day. Fittingly, if you are supposed to save the best for last, it was also one of my final days in this place, if you don't count these dream visits.

I'll be back, and often.

Funny. I can usually count on an agonizing leg cramp by now to ground the flight and send me hurtling back toward the hammock beneath the dead tree, ass-over-lawn-pillow. But I'm aloft again. Was there something on this tour I missed? Something I was supposed to remember or see, and forgot? Do I need to double back? No. The truth is, I'm lost. I can't seem to spy the usual landmarks of the city that are supposed to be sprawled out below. It's so green and sparkling blue — just like Clayoquot.

Why, just over there is the flagpole on the back of my garage, flying the Canadian flag on top and the Tofino flag by Roy Henry Vickers just below. Wait! There's my hammock and me in it! I'm opening my eyes. I'm looking up into the branches of my dead tree and instead of bare limbs, every single arm of this tree teems with life — there sits the Canada geese, the pair of herons, the greater yellow legs, the raven and the eagle, the gulls and the western sandpipers, all looking down at me lying in the hammock. To me, my tree has never looked more vibrant, more nourished, more relevant. I rub the dream from my eyes and realize that I have hung the weight of my very best memories on those branches. *Hi there, guys.*

What a trip.

Never Say Never

– KEVEN DREWS –

My damaged vertebrae ground against each other and a dull pain shot down my spine as I tried to pull myself out of the waist-high water and onto my floating longboard. I felt my puffy, round face wincing; the pain was too much. This wasn't going according to plan.

It was the 2004 August long weekend. For the previous year and five months while I battled cancer in isolated hospital rooms, outpatient clinics and my parents' Surrey, B.C., home, I had prayed for and dreamed about this surfing session: I'd slog into the water, push myself onto my board and, with an arched upper back, paddle effortlessly to the line-up. I'd straddle the board and wait for a set; pick my wave, turn and face the beach; I'd paddle, feel the momentum build, pop up onto my feet and ride. This drill was second nature to me — I'd done it thousands of times. This time, I had to struggle to roll my neoprene wetsuit over my legs and stomach, which were swollen from anti-rejection drugs and steroids. But I was alive and back in the cold black water of Chesterman Beach.

Just metres away in the fickle, knee-high summer swell floated my wife, Yvette, on a bodyboard, and my cousin, Jason, on a surfboard. Yvette stands about five foot seven, has red hair and a slim, but strong, curvy build. Jason is a little taller, has short black hair and a swimmer's build. Next to them was a pack of young, wiry surfers who made everything look so easy — picking, paddling and popping up on the small waves at will. Despite the cloudy sky, everybody was laughing. Meanwhile, I stood in the water, feeling like a marshmallow, cursing under my breath.

"Get it together, Keven," I told myself. "You could do it before, so you can do it again."

I gritted my teeth and slid on. My back jarred. Sweating, I winced again, but I'd made it onto the board, my face resting on its deck. After gathering strength and composure, and with the board's blunt, round nose facing the horizon, I started to paddle. I tried to arch my upper back, and raise my neck and head so I could see where I was going, but my back was too stiff. Cancer had destroyed too many of my vertebrae. This wouldn't do. If I couldn't even paddle, there was no way I'd ever catch a wave. I slid back off the board and stood in the water.

"What's wrong?" Yvette asked.

"I can't do it," I responded. "Let me try the bodyboard."

We swapped boards. *This will be easier*, I thought. I waited for a set of small waves to come. I turned to face the beach and began to kick. Nothing. My legs were too weak and I could barely bend my back. After kicking in futility for a few minutes and catching nothing, I gave up and stumbled out of the surf. I spent the rest of the weekend in bed, on my back, in pain, grinding my teeth. *I have failed.*

Until that day, I never imagined I'd be in Tofino and be unable to surf. After all, Tofino and surfing were one and the same thing for me, and for years, surfing was the reason I visited Tofino. So strong was my obsession that when I graduated from journalism school in 1997, I applied for my first newspaper job at Ucluelet's *Westerly News*. I wasn't sure if I would like the industry, but I knew I loved surfing. The job worked out and my career took off, taking me to daily papers in B.C. and Washington State. I traded surfing for career. I could return anytime to surf — or so I thought.

"Your body's doing something really bad to itself," my doctor said. I lay under coarse white sheets on an ambulance stretcher in the Emergency Room of New Westminster's Royal Columbian Hospital. She spoke of tumours, lesions and bone damage: one vertebra gone, others damaged. What other doctors thought was only a slipped disc a few months earlier was actually much worse. As I took hits of nitrous oxide to control my pain she delivered the kicker: "You've got cancer."

It was March 31, 2003. Just four years after I'd left Tofino — trading surfing for a career — I began to realize there was a chance I'd never surf again. We cannot predict what matters in moments like this, where the mind fixates to process what was happening, what will happen and what might happen. Surfing suddenly became crucial. I'd lost ten centimetres in height, my bones looked like Swiss cheese and I couldn't walk without a cane. Worse yet, doctors diagnosed me with multiple myeloma, a disease in which bone marrow's plasma cells turn malignant, accumulate in the marrow and cause anaemia, bone and kidney damage, immune system problems and infection. The condition masks itself, usually attacking people of predominantly African-American origin who are sixty or older. I was just thirty, of European origin.

To survive, I needed a stem cell transplant. One month after diagnosis, I received radiation treatment. Just weeks later, my brother and I were tested to see if he would be a suitable stem cell donor. Luckily, he was nearly a perfect match, and in June, I was admitted to Vancouver General Hospital for the transplant. For the first six days, doctors administered three drugs — a dose of chemotherapy fatal to all bone marrow — into my body through an IV drip. The chemo killed my marrow and, hopefully, its ability to create cancerous cells. Two days into the transfusion, I was placed in isolation and forbidden to leave. Day seven was a day of

Winter surfing on the west coast, near Tofino. (PHOTO: JEN PUKONEN)

Snowy Chesterman Beach, Tofino. (PHOTO: JEN PUKONEN)

rest, and on June 20, 2003 — which has become my second birthday — doctors transfused my brother's stem cells into me through an IV. We waited for the cells to seed in my bones and create new marrow. While in isolation, I realized this was the first summer in eighteen years that I hadn't surfed.

About a week after my transplant, blood tests proved my brother's cells had seeded and were creating new marrow. Near the middle of July, doctors released me from hospital. It took me fourteen months to gain enough strength to get back into the water, and at least once a day for those fourteen months I dreamed of returning to Tofino — to live, if the transplant succeeded, or to die, if it failed. For months, the latter was a real possibility. When I left hospital, I had no hair, no strength and no immune system. I could have died had I contracted a virus or bacterial infection. I walked with a cane. Surfing was a distant dream.

Recovery would have taken a lot longer had Yvette not motivated me and whipped me back into shape. Just days after my transplant, she took me for a walk at Belcarra Regional Park in Port Moody. If I needed a fix of the beach or ocean, she drove me down to White Rock or Crescent Beach. It wasn't Tofino, but it helped. One week before she began her final teaching practicum, she took me to Bear Creek Park and forced me onto the 400-metre track. I couldn't finish one lap. So, we set a goal: I'd stop using a cane by the time her parents visited from Perth, Western Australia, in December 2003. I tried to walk every day. When Yvette and I decided to visit Tofino a few weeks before her parents arrived, victory came early. I left my cane in my parents' cabin after our third or fourth walk on the beach. It was a small step forward in a tediously long recovery.

After visiting Tofino for eleven years, my parents had purchased the cabin on Lynn Road, Chesterman Beach, in 1988, when land was still affordable. This was home, but I still couldn't be here. Just months after my transplant, the doctors diagnosed me with graft versus host disease, a disease in which the body's new immune system attacks the internal organs. My liver was under attack. Tofino's small hospital was not equipped to deal with me. To fight the disease, I was started on anti-rejection drugs and steroids back in Surrey. I was constantly tired, sleeping up to sixteen hours a night. I awoke every morning feeling sick, as if I had spent the night before drinking heavily. If that wasn't bad enough, our family had to weather break-ins and home invasions. Just days before Yvette's parents arrived, a junkie — HIV and Hepatitis-C positive, suffering from heroin and cocaine psychosis — broke into our house while we were home. He bled all over, and I had no immune system.

About seven months later, a friend and I were enjoying a warm summer's day in my parents' back yard when a man wearing a thick jacket confronted us. Hands in his pockets, he told us he needed $425 for rent. I guess I'd made the mistake of leaving the seven-feet-high chain-link gates open.

"This is no place to recover," I told my parents. "I want to live in Tofino again."

"Let's wait and see," my mom would say. "Maybe . . . when you get stronger."

In September 2004, one month after my unsuccessful surfing attempt, Yvette and I were back in Tofino, this time we hoped for good. There was a need for on-call teachers; unlike in Surrey, work was available immediately. Our savings had been depleted during my recovery. I had not worked in eighteen months, and there was little chance of my going back to work any time soon. Just days after arriving, Yvette worked her first day as a teacher. She also got a second job as a nanny. While Yvette worked, I recovered — waking up late and napping on the couch. Building up my endurance, we walked almost daily on the beach. I was happy to be alive and to be back where I belonged, but every day I saw people doing something I couldn't — surf. I still could not immerse myself in the wild waves of this place, my home. Until I could, I would never fully heal.

Yvette could see my frustration building with my longing to get back into the water. "Maybe you should try bodysurfing again," she said as we walked on North Chesterman on a sunny spring afternoon. I can't remember what day it was because I had stopped writing in my journal, but I remember seeing surfers in the water, and feeling stronger than I'd felt for years — thanks to our long daily walks. I was also tired of critiquing what other surfers were doing, tired of cursing, tired of feeling resentful. I think Yvette was tired of it, too.

Over the next few days, I sewed up Yvette's torn suit and patched the stitch with a rubber sealant. We were now ready to go. We hit the beach on a warm sunny afternoon. Surprisingly, on one of my first attempts, I caught a wave. More importantly, I didn't feel any pain. *I can do it,* I thought. I was regaining my strength. For the next few months, at least two to three days a week, I went bodysurfing no matter what the conditions were — blowing wind or pelting rain. I'd never take bodysurfing for granted.

Months later the wheels in my mind began to turn: Maybe I *could* surf again. Yvette's friends, Karla and her husband Martin, were visiting from Vancouver. A storm was clearing; the sun broke through, the waves were chest-high and untidy. After an hour and a half in the water, I asked Martin if I could use his surfboard. He agreed, and when the water was at my waist, I slid onto the board, lying in the prone position. Small, broken white-water waves hit me in regular intervals. *Wow,* I thought, *this doesn't hurt as much as it did last time.* I couldn't feel my vertebrae grind against each other, and my back didn't jar. I felt stronger. But as I began to paddle, an old problem reared its head: I could barely arch my upper back, or raise my head to see where I was going. My spine was still too stiff. Stubbornly, I kept paddling, trying to duck dive under the coming white-water waves. I couldn't. Again my back was the problem. I gave up, turned the board around and made my way back to shore.

"I'll have the bodyboard back," I told Martin, who was sitting next to Yvette and Karla just above the tide line. "I can't do it. I'll never surf again."

"You don't know that," said Yvette. "Never say never."

August 21, 2005, proved her right. A group of friends had driven up from the Lower Mainland. We headed to Long Beach where the wind was blowing onshore, making the chest-high waves sloppy. We unloaded my Jeep and changed into our wetsuits. Soon, I found myself floating next to Yvette.

"Can I try your board?" I asked her.

She handed me the leash. I fastened it around my right ankle, turned the board's nose to the horizon and slid myself on, lying in the prone position.

No pain.

I tried to arch my back and neck so I could raise my head while I paddled. I was stiff, but arching my upper back and paddling were no longer impossible. Seconds later, I picked a broken white-water wave, turned the board so it was facing the beach, paddled, and felt the momentum push me forward. And then I did it. Struggling, I stumbled to my feet and stood up. For the first time in almost three years, I was riding a wave. Never was now. Yvette was cheering wildly.

When I was a kid, I dreamed about surfing. Then, I dreamed about living here. When I entered university, I dreamed about summer jobs near the ocean. When I finished university, I dreamed about a newspaper career. Then when I was diagnosed with cancer — an event that has become the defining moment of my life — I dreamed of a simple life in Tofino and the chance to surf again. One by one, these dreams have come true. Here. August 21st was the pinnacle, a revelation. I learned never to say never.

Midnight at Catface

- JOANNA STREETLY -

I could have stayed there all night, I think. Breathing deeply, I would have continued to stare around me, knowing that however long I lingered, my eyes would never be able to take it all in: my pores would never be saturated; my ears would still — even now — crave the windy quiet.

Does perfection have a time limit? When does one stop the clock on a sunset? Could a moonlit mountainside ever disappoint?

There's a shushing across my face and the air is sharp with altitude. Falling away from my feet is everything that I love — the lands and waters of Clayoquot Sound, coloured in darkness and lit with silver. The moon is lumpy, fattening for the wane, and the sea is a viscous skin, deceitful in its limpid swirling. I'm tingling as I stand here. From this eyrie, I want to reach out my arms and read the landforms like braille.

Reprinted by permission of the author. Originally published in The Sound Magazine.

I slide down the skirt of Lone Cone Mountain with my eyes, pausing at the glimmer of God's Pocket. On to Matleset Narrows and dreams of its waterfall: frigid green pools braved only in summer; small, snuffling black bears casual on the beaches, and distant wolves brightening an inky fall night. Closer now, past Bedwell, there'd been a morning of such softness that the sky in the water had been indistinguishable from the water in the sky, until the myriad porpoises had slicked up for air all around me. Down there, past the landslide was the pygmy owl in broad daylight, and over there, on Morfee, the cliff must nearly be yellow again with monkey flowers.

These shapes hum to me through the bright night. It is an indescribable language of personal connection and memory — the language of home.

I could have stayed there longer, I know. I could have watched the moon until it set. I could have stayed beyond the moment of perfection, whenever that would be. But it is better like this, I think, because now, as I gaze up at the mountain, I can feel the allure drawing me — enticing me. And I want to go there again.

Love Song to Clayoquot Sound

- SHERRY MERK -

The wild shores of Clayoquot Sound sang a siren song to me for years before I journeyed there, before I ever saw the perfection of its beauty. This song captured my imagination, my heart and my spirit, drawing me to it as surely as a murrelet is drawn to its nest. Now, years after a glorious decade in that place, the home of my spirit, I still hear its call. Always the wild shores of the Sound pull at me, echo through my heart, here in my home-away-from-home, where I am remembering, remembering. It is a song of the clean and pungent air, the salt spray, the sea foam, whitecaps lined up and galloping in to shore like wild white horses, manes a-flying in the song of the wild waves, rolling and crashing over black volcanic rock, swirling the maelstrom of what is known to locals as "the Cauldron" at the base of the cliffs on Frank Island. This song sings through the forest floor, through the tops of giant cedars, in the lift of the eagle's wing, with seabirds wheeling free above shining waters. It sings inside me still. It always will.

Half a lifetime ago, at age thirty, I discovered I was an ocean person, a *displaced* ocean person, cast up like a beached whale on the unlikely shores of Okanagan Lake. The peaceful lake with its lapping ripples, fragrant bulrushes and weeping willows was pretty, and desirable to many. But it was the wrong scenery, the wrong place, the wrong "energy" for me. I don't just love nature — I have a spiritual and physical *need* to live immersed in it. When I am not, I am homesick, heartsick. When that most essential component to my well-being is lacking, when it is citified, tamed or domesticated, I must make do, and am always aware that I am making do. In a perfect world, I would carve myself a dwelling inside a huge old tree, on the edge of the sea, with only the wild in sight, no human footprint visible but mine.

By forty-two I had been longing for the ocean for so many years that I began to feel my soul dying. I had waited patiently as my three older children completed high school. With my youngest leaving elementary school, about to enter high school herself, I knew it was now or never. I could not face four more years of marking time, as my years and energy slipped away. I grew plaintive about it. I keened, I yearned. Not brave enough to make this large a leap myself, I waited for the universe to make the change for me. And, in its way, it did.

My sister stacked the deck. For my birthday that year she took me to Tofino on a zodiac expedition to see the whales, and it was as perfect an experience as it could possibly be. The sea was serene, there were whales everywhere, the day was clear and sharply etched and, when we turned off the motor, we were drifting on the same level as the whales. They were so unconcerned with our presence that one surfaced beside the boat, thrilling me to my toes. Its ancient eye looked upon us; we gazed back in silent awe, the gust of its every exhalation sounding like the very breath of God.

We passed rocks covered with barking sea lions. Beneath a huge nest in a tall snag we sat, staring at the resident eagle, who stared diffidently back. Little orange-beaked puffins bobbed diligently atop the waves. We investigated little inlets, discovered waterfalls, and as we headed back to shore, sunset spread its palette of colour before us.

The tour guide spoke of saving Clayoquot Sound's forests, heavily under siege by the multinationals. She informed us that this one remaining pristine ecosystem, this last stand of old growth was being cut and exported as raw logs — much of it to be pulped and made into phone books for California. Logging was spreading towards the few remaining untouched watersheds. It was time for those who cared to take a stand. My heart caught fire. Everything I loved, longed for and believed in was here, and I wondered, why am I not here too?

I did not want to go home. I did not want to be a tourist who had to leave. I wanted to sit by local activists' fires, join in their talk, and be one of them. With every fibre in me I knew that I belonged here. But I went back to the Okanagan, to my hated job and to a persistent depression as winter closed in. After a few weeks, I wrote a letter to the tourboat owner who had guided our excursion. I told her how lucky she was to be living her dream, and how long the west coast had been my dream. On our trip she had spoken of her difficulty finding anyone who could handle her business as well as she did, so that she could take some time off. In my letter, I ventured to wonder if I might be that person, if there might be a place for me there.

There was no reply; the winter went on. The walls closed in and I felt trapped by the need, as a single mother of four, to earn a living, to support the kids, by my aloneness and the seemingly endless struggle. I felt as if I existed only to bring brown paper bags of groceries in the front door. An aware employer recognized my spirit was faltering. She encouraged me to take supervisory training and apply for the position of supervisor in another department, away from shift work, into management. I passed the training, won the position and for the first time was earning enough money to pay the bills. But the universe, having a sense of humour, chose that moment to offer me in the form of a letter from Tofino, the choice it has always presented: continued "security" (a huge issue for a single mother accustomed to poverty) in the job I had recently been given, or the life of my dreams, and utter insecurity — part-time work at six dollars an hour, but in Tofino where I longed to be.

I wrestled with the enormity of the choice, with its uncertainties and all of its unknowns, but there was little doubt. I knew this was a choice about following my heart, or giving up my dream and staying where my spirit was dying.

Had I known the difficulties this choice entailed, I might never have found the courage to make the leap. I am glad I did not know. It was the biggest risk and leap of faith I had ever taken, and it repaid me with ten years of unparalleled joy.

The night I rounded the corner at Long Beach in the rented Budget truck, a gigantic ball of fiery scarlet was going down behind the hills, the sky a Gauguin canvas. Taking a moment from unloading boxes into my winter cabin, I saw a whale in the bay — a whale in my front yard. From the moment I first set foot on the beach, that questing, longing, seeking voice inside of me was stilled. I was at home, the home of my spirit, the one place on the planet where I belonged. No matter how difficult it was to find housing (a merry-go-round repeated every few months), or how many part-time jobs it took to pay the rent (two or three on average), whatever it took I would do to feel this rightness, this boundless joy, this sense of being exactly where I was meant to be. Home.

For the next ten years, I walked ecstatically through some of the most spectacularly beautiful landscapes on Earth, with daily joy and gratitude at the visual feast, and a fullness in my heart that meant more than any amount of money. Happiness abounded, free for the taking, free for anyone who had the eyes to see.

That first winter, I lived in a wooden cabin on Chesterman Beach. Frank Island was across the sandbar out my kitchen window. Every morning, as I plugged in my teakettle, I caught my breath in wonder, looking out on white-topped waves, a scene of perfect and unimagined beauty, mine to look upon: *mine!* My eyes loved everything they fell upon: treetops poking out of the early morning fog; beckoning miles of white sand stretching to infinity; herons picky-toeing along the pebbled mudflats; orcas vaulting by. Lone Cone blushing deep rose at sunset; little boats all heading into port through the dancing waves; interestingly attired alternative lifestyle folk drifting serenely past on bicycles, sometimes singing; the cute little village centre, its appearance unchanged even now, after ten years of a hundred million visitors clambering over it like ants on an unruly anthill.

The sights of home sang a constant love song inside me. Rain-slicked, gum-booted locals, heads bent against the wind, making their way laboriously to the post office under the lashing rains of winter. The bell tinkling on the door of the Common Loaf Bakery as I went in out of the rain, local faces upturned to greet me, wet rain gear sloughed off and steaming. Radar Hill on an early springtime morning was the perfect beauty of the natural world for 360 degrees. The sounding of the foghorn at Lennard Light all night long through the eerie, drifting fog. And in the morning, it's the complete silence that lets you know the power is out again, no humming appliances to muffle the sound of perfect peace.

There was the Tall Tree Trail on Meares Island, where I would visit the Hanging Garden Tree, an opulence of ferns and smaller trees cascading from it, and the Stairway to Heaven tree, a giant fallen uphill, so you could walk up it, like a ladder to the stars. I had only to step out onto the beach to achieve a meditative state, all worries falling from thought like weathered shakes from a tinder-dry frame. Every day I chose a different beach, an undiscovered trail. I explored every beloved inch of my new home.

❧

When I moved to Tofino in 1989, the environmental movement was gathering steam. This was reflected in the graffiti painted on concrete abutments rimming the sharp corners and steep drop-offs along the mountainous Pacific Rim Highway. We village folk travelled the highway frequently, for shopping, visiting the Big Smoke, appointments with specialists. Heading out, we might read "Hug a Tree" in huge black letters crooked against the concrete. Coming home, on the neighbouring abutment, in even bigger letters we might see "Hug a Logger, You'll Never Go Back to Trees!"

As the environmentalists grew more numerous, vocal and visible, the Share B.C. movement rose in response. "Share B.C." said the pins on the lapels of the ladies at the Post Office, who were married to loggers. "Share the Stumps!" cheerfully responded the greens. "Logging Feeds My Family" proclaimed the high-up bumpers of trucks. "No jobs on a dead planet" said the signs at the blockades.

Then came 1993, and the Peace Camp, and everything came to a head: arrests, placards, banners across the highway ("No More Token Groves" at Cathedral Grove). Those early mornings on the blockades were the most meaningful and fulfilling moments of my life. I was part of this cause and I stood there with my whole heart bursting, especially as part of the Women's Blockade. Proud, joyous, strong, we sang and spiral-danced around the roadway, feeling our power, sharing smiles. My son phoned that night, "Mom, I saw you on the news, singing and dancing on the road with a bunch of hippies!" I remember the gentle sleepy beat of the tom-toms in the early morning half-light as we gathered around the campfire. The fear and determination as the big trucks rolled in, as the official read out the order to move off the road, and the RCMP moved in to make the arrests. People were often carried away, held by their arms and legs, to our supportive cheers and tears, the most profound and passionate hours and days of my life.

I was there the night they closed down the Peace Camp. Dana Lyons sang "Magic" as activists who had worked hard all summer danced in a clear-cut. We shared a fifteen-minute group hug and a long "OMMMMMMMMMM" that ended with blissful smiles amid the stumps under a full, round grandmother moon. I had just missed the hippy movement of the sixties. Now I had this, the best of all. I had come full circle. I had come home.

Offering more than its beauty, Tofino was the *community* that fulfilled my sense of belonging. Some locals believe that ley lines intersect near Tofino in a way that makes it one of the power spots on the planet, that it draws certain folk to it like a magnet, as I had been drawn. It was here that highly original, individualistic, creative, intelligent, authentic, interesting and fully alive people, each with a unique gift to share, enveloped me. I learned to be on "Tofino time." There appeared to be no barriers. At all events, all ages were present, each person accepted as a being of worth. On International Women's Day a potluck was held. Lush young women at the peak of their flowering, demure young girls just on the threshold, grey-haired and decorous-enough-looking crones (a term we reclaimed as positive) — all chatted together until the drums began, at which time we were all transported into a spiral-dancing, writhing, joyously beaming mass of primal womanhood. A beloved and feisty septuagenarian got up to do a boogie-shimmy to our raucous delight. A young woman read a poem about menstruation that had the room in hysterics. A belly dance was performed with the knowing eyes and awesome sensuality of a full-figured woman at home in her body, followed by a slender First Nations girl draped in white wolf fur who danced for the animals.

After 1993, the tourists came in the millions to see the place that had gained so much attention on the evening news. The tourist dollars pouring in changed some of the feeling and the flavour of our little village. In terms of quality of life for the locals, business and development decisions began to be a lot more about money than many inhabitants might prefer, especially the low-income ones. Everything started tuning up to welcome the new money, and graffiti no longer fit the image encouraged by the moneyed interests. The abutments returned to pristine unlettered concrete. Some of the loggers began to understand that their livelihoods were more at risk through forest practice (mis)management than from the efforts of the eco-warriors. Ultimately, their jobs disappeared and Ucluelet had to turn its focus to tourism as well. The banners were folded up and put away, and I had to move to Port Alberni. Somehow there isn't a bumper sticker that feels right for my life here. For those who guide the decisions, the priority is continuing the upscale trend that is "Whistlerizing" Tofino, effectively squeezing out some of the people who love the place most dearly.

Why I had to leave Tofino is another story. I worked at too many jobs, which led me to a collapse that eventually cost me my place. No choice but to sell my hard-won trailer, my tiny foothold on this place so dear, I was forced away to the west coast's nearest neighbour, chosen because it is as close to "Tough City," our local name for Tofino, as I can afford to live. I left Tofino suddenly, by ambulance, with no time for a painful goodbye. I had to learn to be happy in yet another landscape foreign to my spirit. The universe set me the task of losing, once again, what I most love, of learning to be happy — if not ecstatic — in the least likely setting I could envision for myself. Is it better to find home, then lose it, than never to have found it? Yes, definitely.

For ten years, I was so in love with Clayoquot Sound that I had no need of a partner. How many lives can claim a solid decade of joy on a daily basis? I was in the august ecstatic company of Rumi and Hafiz. My soul sang with one constant refrain: *this place is so beautiful, I am so happy to be here.* And *I did it myself* — with generous assists from all the gods and angels, who assist me still. I miss it every day, the dream of returning a beacon onto which I fasten my hope. I no longer have health, finances or stamina to re-enter the housing merry-go-round, but one day, perhaps I will return. I want to end my life there, where I belong, among the loved and venerable crones in that eccentric-friendly place. I want my remains put to rest in the cemetery under the wind-whipped pines, where the winter gales howl through. Love for that place never stops singing inside me. I carry those ten years within me like a gleaming treasure, like a song of love.

Immersing

An Excluded Sort of Place

– SUSAN MUSGRAVE –

As I crumple an old newspaper to light a fire in the wood stove, an item catches my eye. Every day a teacher, Yang Zhengxue, treks an hour and a half up and down craggy peaks in a remote corner of southern China to reach his students who live high on a mountain ridge in a limestone cave, "an almost prehistoric habitat without electricity, running water or any other amenity that would identify it as a home for residents of the 21st century."

Sounds like home sweet home to me. The newspaper is dated December 11, 1983, but the news hasn't changed much. Weapons of mass misery, genocide, political blunderings — the so-called real world seems a long way from these misty, mystical isles. Here, in the wilderness, I encounter a different sense of time, where my days

---

Reprinted by permission of the author. Originally published in You're in Canada Now . . . A Memoir of Sorts *(Thistledown, 2005).*

fail to follow orderly paths, unfolding instead in unpredictable ways. I reset my body and mind to a cosmic time frame, plan my activities around the incoming and outgoing tides, the rising and setting sun.

There is no such concept as *mañana* on Haida Gwaii; no word exists for that kind of urgency. A tourist once stopped for Buddy — a local character who spent his days walking the five miles, back and forth, between Skidegate and Queen Charlotte City — and asked him if he needed a ride. "No thanks," Buddy replied, "I'm in a hurry."

There's no cellphone service here, no schedules to keep. Pockets of these islands missed the last ice age, but a few of my technological pioneering friends have recently joined the 21st century and bought transistor radios, so they can find out the current time by tuning in to the CBC.

A few summers ago when we visited Jim Fulton (who had been the NDP MP for the Skeena district for fifteen years) at his log house on the Tlell River, miles away from any "conveniences," we sat around the wood cook stove eating deer meat by gaslight. "Have you ever heard of electricity?" my citified daughter, then aged six, asked the Fultons' kids. "Well, you should *get* it!"

Truth moves to the heart as slowly as a glacier, and that's how time moves here. I've been waiting at the side of the road with a garbage bag full of dirty laundry for most of the morning (my truck has a dead battery); the first fisherman to come by in his half-ton, stops. I settle into a pile of gill nets on the seat beside him, then ask if he has the time.

"I used to wear a watch," he says, "but I lost it in the winch." He shows me an empty sleeve. "I know one thing for sure. There'll be a high tide tonight."

The fisherman — Crabby Mike, who once rode with Sonny Barger and the Hell's Angels — came here twenty years ago. He sailed

Moresby Camp, Haida Gwaii. (PHOTO: JEN PUKONEN)

SGang Gwaay, on the exposed southwest coast of Gwaii Haanas, the UNESCO World Heritage Site, also known as Anthony Island. (PHOTO: JEN PUKONEN)

his boat up and anchored it in Masset Inlet, where it started to sink.

Mike rescued his fishing rod from below decks, and sat on the tipping deck, drinking a Bud Light, fishing. "I'd come to the Charlottes to fish," he told everyone, afterwards, "and that's just what I intended to do. Fish."

People from all walks of life come here for the fish. Our former prime minister visited a couple of summers ago, too, but I doubt whether Monsieur Chrétien had time to visit the Best Little Lure House in the Charlottes, in Queen Charlotte City. He certainly didn't venture into Haida Bucks in Masset for a latté; he ate lunch aboard his airplane on the Masset landing strip.

A school bus full of llamas, their heads poking out the open windows, draws up in front of the laundromat, where Crabby Mike deposits me. Nobody gives the llamas a second look — they're used to it, just as people are used to living with wild and unpredictable weather.

"But doesn't it rain *all* the time up there?" a friend, planning a visit from Vancouver, writes. (Okay, so we're not completely cut off: there *is* a daily mail service — providing the plane from Vancouver can land, on a runway stretching between two driftwood littered beaches at the Sandspit airport.) And yes, the rain it raineth. The rain falls so hard here it bounces off the ground then goes back up.

It's no surprise that the windshield wipers on my pickup have a life of their own. Even if the rain lets up, and I reach to switch them off, they don't miss a beat.

There'd be no point trying to get the wipers fixed. Once when my father came to visit from Victoria, we ended up having the inevitable discussion about "island time." Dad said even on Vancouver Island these days you couldn't expect to get anything done in a hurry. He had taken his radio to an electrician two weeks ago and he still hadn't got it fixed. "That's nothing," said my elderly neighbour, Frieda Unsworth. "I took my car to a mechanic in Masset seven years ago and I haven't got it back yet."

❧

Masset's the kind of small town where you recognize everyone by the vehicle they drive. The liquor store employees can tell you what anybody in town drinks, and whether they've been drinking too much, lately, for their own good. A trucker I met laughed when I mentioned that the Masset Liquor Store must be the only liquor outlet in the world where visitors are asked to sign a guest book as they leave. "Sometimes I sign it twice a day myself," he said. (Comments in the Guest Book range from "excellent liquor" and "Drunk, 7-24-98 10:23 A.M. Wolf Parnell" to "The wine here is cheaper than the gas.")

When people find out where I live they often ask, "Don't you feel isolated, living out there, away from it all?" I even had a taxi driver in Toronto ask me if I came from an "excluded" sort of place.

Living in seclusion, or exclusion if you like, does have advantages. People are forced, by circumstances, to be polite to one another. You can't risk running the local undertaker off the road when he turns in front of you without signalling: you may need his expertise one day.

Chances are you know where he's headed, even if he forgets to indicate. Living in a small town you never *have* to use your turn signals: everyone knows where you are going, anyway.

In Masset, everybody knows where you've *been*, also. They know what you do for a living, and who you are married to — at the moment. For this reason I always try to behave myself when I go to town. I'm especially well-behaved in the post office, because mail is my livelihood, and the last person I want to alienate is the person who sorts my cheques. A friend who got a job sorting mail at Tlell found a registered letter for an island poet, Hibby Gren, who had been dead for ten years. The postmistress, back then, had decided Hibby had been too drunk to be entrusted with important registered mail, so she'd held on to it for fifteen years.

At the north end of the world, home to any number of social misfits who have fled from the normal stresses of 21st-century living, the barter system is alive and well. If you want *all* your garbage hauled away, you leave a reefer on top of your can as an incentive for the local collector. In the old days when they hired strippers at the hotel, one famous North Beach comber is said to have offered an exotic dancer fifty pounds of shrimp to pass the night with him.

Life is simple, here, pared down to the bare necessities. There is only one traffic light, for instance, and few other signs giving directions. There *is* a sign announcing WIGGINS ROAD AHEAD, a few metres before you come to Wiggins Road itself, and that tells you something about the frequency of side roads off the highway connecting Queen Charlotte City with Masset. And then there's the WILDLIFE VIEWING sign underneath the one indicating "Masset Cemetery Road." From what I remember of the people who are now permanent residents in the graveyard, "wild life" viewing would be an understatement.

The graveyard is one of the places I like to spend quiet time, though. The trees drip moss, the graves themselves are overgrown with salal, salmonberry bushes and more moss. I haven't chosen the precise spot, but one day I expect to take up residence there myself. "Everybody loves that graveyard," said my friend Henry White, as I helped him undecorate his Christmas tree on the Ides of March. "One guy even *hanged* himself in there."

The simple life gets busy here. Between cleaning crabs, gutting deer, canning razor clams and picking huckleberries for pie, I finally find time to finish the article about Yang Zhengxue and the limestone cave dwellers. The only modern appliance they possess is a battery-operated red plastic alarm clock. Twice a day its owner lets it chirp on and on, for forty-five minutes or more — "not to tell the time, but to entertain, like music from a Stone Age radio."

Those cave dwellers sound too high-tech for me. And no place else, either, compares to living where can you see seven rainbows in the sky at once, or count twenty-five eagles perched on the same rock, or lie out under the stars in August and watch the Perseids, the northern lights *and* forked lightning all at the same time. When the storms start howling and the plane doesn't come in, and the ferry is stuck in Prince Rupert and supplies are running low in the Co-op and the Government Liquor Store, there's no place on this earth I'd rather be.

November Day of the Dead

Seeking Y'aq-wii-itq-quu?as (Those Who Were Here Before)

– DAVID PITT-BROOKE –

I explore the point and the nearby beaches until it's time to head home. I don't plan to linger. The sun is still shining brightly but I'd like to reach home while the weather holds, preferably in daylight. It's been a good trip, a happy trip, and I want to keep it that way. It's a successful trip too: I have a much better sense of what life might have been like for the people who lived here, my goal for the day accomplished. Even so, I can't help feeling a faint disappointment. I was hoping, against all reason, for some clear sign of human habitation, something to validate the site, a direct link to those people who went before. I know that I'm chasing a will-o'-the-wisp, but I can't help it. It would be nice to have something tangible, something I could touch with my hands.

Excerpt from Chasing Clayoquot: A Wilderness Almanac *(2004). Reprinted by permission of Raincoast Books.*

I'm just starting back, making my way cautiously along the top of a drift log, one of an immense pile at the head of the beach, when I notice a game trail going off into the bush, exceptionally well-worn, with just the suggestion of an opening back there. I'm a sucker for these secret hidden places. Perhaps there's a little marsh back there, a lagoon or a little sandy dune, a nice camping spot for some future trip. Probably it's nothing at all. I've been fooled many times, seduced by likely looking trails to nowhere. I'm anxious to be on my way; I'm already thinking about the kayak. But there is something about this trail that intrigues me, something inviting. I check my watch. Five minutes, I tell myself, as I step into the bush.

What I find, when I finally pull free from the salal, amazes me. A grassy clearing scattered with clumps of spruce stretches back a hundred metres or more, park-like, in a series of low rolling terraces, five to ten metres high. There are one, two, three of these, rising back into the forest, one behind the other. Sunshine illuminates the grass. The whole meadow is a brilliant green. A quiet, peaceful place.

This is how it must feel to stumble across a lost city or a forgotten temple, I'm thinking. A large grassy clearing amid the tangle and riot of rainforest vegetation on the west coast of Vancouver Island is an extraordinary anomaly: it's like coming upon a patch of jungle in the middle of the desert or a tropical beach in the arctic: a bit eerie, mystical, certainly haunted. I feel more than a hint of the forbidden world here.

The five-minute limit forgotten, I wander into the open for a better look. Where the meadow finally gives way to forest, the terraces continue, disappearing under the second growth. It's a most impressive site, though far from undisturbed, as I soon discover. The forest all around was logged some time ago. There are deep wheel ruts through the clearing and wide trenches where roadways were cut through the terraces.

These terraces are no work of nature; they're huge middens. Here the debris and refuse of this village accumulated over hundreds or perhaps thousands of years. In places, the grass and moss have fallen away from the steep banks of the road-trenches to reveal the typical ink-black soil, greasy with organic matter, and shot through with little bits of sea-shell, so characteristic of west coast middens. These particular middens have become so large they're interfering with the natural drainage. Small patches of marsh have developed in the hollows between them.

When this place was inhabited, houses stood in rows along the top of each midden-terrace. I look but cannot see any timbers remaining. This is no surprise. Untreated wood does not last long in this climate. But at regular intervals along the top of each midden are clumps of young Sitka spruce. As I approach more closely, I realize that each little cluster is laid out according to a distinct rectangular pattern, very similar from clump to clump; these trees are nursing upon the remains of corner posts and ridgepoles.

It's not hard to imagine the houses as they were — the great posts and beams, huge logs, some freshly adzed, the smell pungent, and the beautiful planks of red cedar — along with the movement and life around the village. In summertime, one might see fishermen unloading halibut from their canoes down on the beach; or a group of women going off with burden baskets on their backs to gather berries or maybe eelgrass roots; or a canoe being paddled through the reefs offshore. At this time of year, on a day like today, there would be people outside enjoying the break in the weather. Women might even bring their weaving or basketry outdoors. At the far end of the beach, there might be a couple of men using the sunny afternoon to put final touches on a new canoe: pitching the seams perhaps, or scorching the hull to smooth it and remove the slivers.

I sit on a large branch that grows low along the ground from one of the corner-post spruces, sticky with pitch but comfortable. The view down the meadow is especially lovely. A small airborne column of insects dances in the sunlight nearby. Where do they come from at this time of year, I wonder? They must tough it out all through the rainy days, waiting and hoping for the sun to shine, just like the rest of us.

The typical historic site focuses on a brief era, sometimes just a single event, a crisis, an extraordinary but passing phenomenon. But in this place we receive a very different perspective. Here is an extraordinary accumulation of ordinary life, daily events repeated thousands upon thousands of times, over thousands and thousands of years, gradually accumulating like layers of earth, like the rings of a tree. That is what speaks to us. Even now, I can feel the weight of all that humanity, a great multitude if they could but gather in one time and place. The meadow is practically humming with human presence, slightly muted, as if felt from a great distance. It is not alarming, not scary. I have a sense, you know, of being *welcome*. Perhaps they're pleased that I've come to visit; I don't suppose anyone else has dropped by for a good long time. It's a beautiful spot. I don't want to leave. It occurs to me that I have, in a minute way, become part of this community, part of humanity's relationship with this place, by simply finding my way here today. The thought pleases me very much.

Who were these people: hundreds of generations, thousands of years? Nobody knows. Nobody can know. There is no record. The only message they've left us, the sign and symbol of their passing, is in the accumulated earth of these great middens. But I feel a powerful sense of kinship nevertheless. Perhaps that is what gives this place such significance. A local site like this echoes a specific local aboriginal culture, certainly, but it also speaks more broadly of simple human endurance under difficult circumstances. In that sense, all humanity has a stake in what happened here. This is part of the story of our species, testament to the everyday heroism, strength, tenacity of which human beings are capable, holding this storm-swept point at the edge of the north Pacific for millennia, a source of pride for us all.

Places like this also echo a tradition, a way of life, that, in its finest elements, is not peculiar to one First Nation or another but is a reflection of the country itself — a quiet, simple life between ocean and forest on the wild west coast, a life lived in a fair degree of harmony with the natural world. It is a tradition that many still aspire to, among the First Nations and everyone else. In that sense, the people who went before — the Y'aq-wii-itq-quuʔas — are forebears, ancestors, to us all: even to those who may have migrated rather more recently than the ice ages.

Too late, as I sit here, it occurs to me that I'm probably trespassing, not against those who went before but against the present-day owners. A place like this has got to be a native reserve, surely, private property. I respect that ownership and regret the intrusion. Even so, I'm not sorry to have seen it. Maybe I even have some sort of modest right. Somebody else owns this property, no argument. But — with all due respect to a people who have every reason to suspect the acquisitiveness of newcomers — I'm not persuaded that they own the past it represents. That belongs to every one of us as we take our place in this story.

Along the furthest boundary of the meadow is a line of very large stumps, perhaps a memento of the drive for Sitka spruce wood during World War II. Wouldn't it have been something to see them towering over the village? I'm fifty years too late. But then I discover — down at the very end of the line — a living giant. For a moment, I wonder how it escaped. Then I look up and discover that the main spar is broken off not too far above the ground, casualty to some ancient storm. Deformity saved this tree's life; it wasn't worth taking. Around the snag, the living branches have turned upward to become trunks in their own right.

The main trunk — I reach out and touch it — has a diameter of at least eight feet, maybe ten. It's far larger than anything growing down on the meadow; must be four or five hundred years old, a witness to daily events all through that time, my tangible link with the past. I listen respectfully, just in case this tree has anything particular to tell me on this day. Nothing comes through, but no matter. For me, it is enough to be here. I'm content with the day, content with myself. All is quiet and peaceful in the sunshine. The sound of surf is muted. I stand next to the tree and look down across the village. I think about what this tree has seen here, watching patiently day after day, year after year, century after century. Above all, I feel the comfort in this place. More than any other experience in this year of rediscovery, I think, this abandoned village embodies the essence of what I was chasing when I came to the west coast of Vancouver Island all those years ago. Perhaps I have come home at last.

From the Heart of Clayoquot Sound

- ELI ENNS -

There is a place here in Clayoquot Sound where two sharply distinct areas meet. At a place called Clayoqua located at the confluence area of the Clayoquot River and Cold Creek, pristine ancient-cedar forest contrasts with recovering, clear-cut riverbanks. Clayoqua is at the crux of what once was, and what has come to be. It is the place where recent human history in Clayoquot Sound begins; it is the place where Tla-o-qui-aht come from; it is the heart of Clayoquot Sound.

My name is Eli Bliss Enns. My mother is Karin Enns, of Peter Enns and the late Dora Enns. Peter and Dora are first and second generation immigrants from Europe. My father is Albert Charlie Jr., of the late Albert Charlie Sr., who was the third born son to the late Harold Charlie, better known here in Clayoquot Sound as Now-waas-um. Now-waas-um's parents were Tsee-iism and Quatia; they were of the last generation of Tla-o-qui-aht who did not have

English names. Now-waas-um was the official historian and speaker for Wickaninnish, the highest ranking chief of the Tla-o-qui-aht confederation.

It is a long-standing custom to introduce oneself in this manner when addressing others in a formal way. It is important to know who you are, and where you speak from, and it's important to convey that to others so that they understand the context of your words. It's also a basis for truth. Our ideas grow from this foundation, and in a sense this is how we qualify the words that come afterward. In other words, this is who I am and this is my truth.

In a basic sense, I consider myself and all of my relations to be Quu-us. Quu-us is a term that we use to describe ourselves. It doesn't translate to mean First Nations or aboriginal, and it doesn't mean white, black or brown. Quu-us means "real human being." We are all real human beings. This is the starting point, and this is our common ground with all other peoples.

TLA-O-QUI-AHT: A BRIEF INTRODUCTION

Tla-o-qui-aht is the confederation of historic native groups that once lived all around the lake system called Ha-ooke-min. Tla-o-qui-aht has been translated to mean "different people." However, it means much more than that. To begin with, Aht means people, and Tla-o-qui is a place in Clayoquot Sound presently known as Clayoqua. In this way Tla-o-qui-aht can be understood to mean the "people from Clayoqua."

This understanding of Tla-o-qui-aht speaks of the history of our people dating back to the early to mid-1600s.[1] As mentioned, in former times, our ancestors were in fact not one tribe, but many small tribes and family groups who lived all around Ha-ooke-min, which is now known as Kennedy Lake and which is where Tla-o-qui is located.

The defining event that changed the face of Tla-o-qui-aht forever is eternalized in the name of the Esowista Peninsula.[2] The war of Esowista was the first great war that Tla-o-qui-aht engaged in as a single force. The people who once lived on the peninsula from Long Beach to Tofino and further north had kept tight control of ocean resources, and had made it a common practice to raid the sleepy fishing villages of Ha-ooke-min to take slaves and other commodities. In our language Esowista means "clubbed to death."

Tla-o-qui-aht maintained their presence in this part of the Sound through to first contact with Europeans in the late 18th century. Having already been engaged in trade with neighbouring communities, the extension of this practice to the newcomers was a natural transition. For the most part, trading relations were peaceful until the early 1790s when an American trading vessel called the *Columbia* fired on Tla-o-qui-aht's principal village Opitsat.[3]

In summary, Tla-o-qui-aht, different people, are the people from Tla-o-qui; they are a confederation of many different smaller groups who once lived a very different lifestyle at Ha-ooke-min.

CONFLICT WITH THE NEWCOMERS & THE STORY OF THE *TONQUIN*

Having recently gained independence from Great Britain in 1783, the United States at that time was a young industrious country whose leaders had adopted an expansionist policy (later given the name of manifest destiny) into their own foreign policy.[4] Testing its borders to the northwest, the US was involved in an escalating competition with the British colonies of the north for control of the lucrative trade that was found in the Columbia Country. The Columbia Country at that time consisted of what is now Oregon, Washington State, the mainland of what is now British Columbia and Vancouver Island. In 1810, control of most of this territory still rested with the sovereign indigenous nations who had lived there since time immemorial. Wickaninnish of Tla-o-qui-aht was well known to the newcomers as "the emperor of the sea coast."[5]

With the burning of Opitsat branded into his memory as a defining moment of his young life, Wickaninnish was bound and determined to establish his own fleet of armed ships to protect his people from the newcomers and to maintain his sovereignty. These desires led Wickaninnish to arrange for the purchase of a ship sometime in the early 1800s, but when the deal fell through, Tla-o-qui-aht turned to a planned attack aimed at taking a ship by force.

The unlucky ship that became the target of this plan was the *Tonquin* of John Jacob Astor's Pacific Fur Company.[6] Setting out from New York in 1810, the *Tonquin* established Fort Astoria at the mouth of the Columbia River four months before the arrival of an overland expedition of British traders from Montreal. Establishing the fort was a major success for the Americans as it secured a strategic command of the Columbia Country.[7]

The expedition was then scheduled to make its way up to Sitka, which was then a part of Russian North America, with a brief stopover in Clayoquot Sound before heading to Asia to offload its merchandise before finally returning to Fort Astoria. However, things did not go as planned. Long days turning to long nights of prayer brought the ship into Clayoquot waters. Our warriors attacked after their sub-chief was insulted by the captain. The very last two crew members undertook a suicide mission, setting the cargo of gunpowder alight. An explosion resulted such as had not been heard in Clayoquot Sound before or since.[8] Many of our people died, and the ship eventually sank. To this day, only an anchor and glass trading beads have been found, and even these underwater archaeological treasures cannot be proven to be from the *Tonquin.*

As a child, I clearly remember walking down to Tonquin beach and feeling as though I were in the land of giants. The trees towered high above me, the ocean stretched out over the world out of sight and the mountains behind me seemed to prop up the big sky.

The closing of the fur trade wars with the Nuu-chah-nulth in 1811 was followed by the brief war of 1812 fought around the Great Lakes between the British colonies of the north and the United States. This occurrence was the first of a string of events that ultimately led to the establishment of the forty-ninth parallel as the border between the United States and what was soon to become the Dominion of Canada. Tla-o-qui-aht were largely unaffected by these developments.

Sea foam on Chesterman Beach, Tofino. (PHOTO: MARKUS PUKONEN)

Dug-out canoe, Flores Island, Clayoquot Sound. (PHOTO: JEN PUKONEN)

CHANGE

In 1871 the colony of British Columbia joined the Dominion of Canada — without any consultation with Tla-o-qui-aht. At this time, relations with these new governments were carried on largely through the church and the Canadian military. There were divisions in our communities over how best to deal with the ever-growing presence of European culture and customs. Even as these debates were occurring, however, the biggest change — devastation of Tla-o-qui-aht social structures through the institutionalization of our children — was underway.

At the same time, the Canadian military was making inroads into the Tla-o-qui-aht hereditary governance system. The relationship that developed is evidenced in alliances between Tla-o-qui-aht and Canada during the Second World War. Following the Japanese attack on Pearl Harbor, Tla-o-qui-aht cooperated with the Department of National Defence to build an air base at Esowista, along with several anti-aircraft gun emplacements, bomb shelters, and barricades against landing craft along Long Beach. Additionally, a high-ranking hereditary figure within Wickaninnish's house enrolled his second eldest son in the Canadian Army. Upon returning home from Europe with a medal, that individual became the first elected Chief Councillor of the Clayoquot Indian Band.

Throughout this time, the relationship that our people had with the land was beginning to change as well. Following the Indian Act of 1876, and subsequent provincial legislation that was applied to traditional lands, Tla-o-qui-aht slowly became alienated from our lands and waters.

Many people coming out of the residential schools got work on fishing boats or made their own boats to take advantage of the lucrative industry. Some worked in logging camps and learned to see their ancestral lands through the lens of federal and provincial jurisdictions, including tree farm licenses and international and domestic waters designations.

Some of those who came through this system saw first-hand the destruction that primary industry was having on the land and on aquatic resources. With a message of urgency from the elders and the hereditary guardians of the land, Tla-o-qui-aht joined forces with other local groups of concerned Quu-us in the early 1980s to stop the planned logging of Meares Island.

These actions spurred on the creation of the British Columbia Treaty Commission, and gave new hope to First Nations in B.C. who had been having their land claims largely ignored for 130 years. It also gave new hope to those who saw a different potential for the lands and waters of Clayoquot Sound.

THE WAY WE SEE THE LAND, AND THE WAY WE SEE OURSELVES

A discussion of self and the meaning associated with the land go hand-in-hand in Tla-o-qui-aht culture. I hesitate to use the word "culture" because I was told by one of our old people that there is no word in our language for that term. According to this fluent speaker, the best way to describe culture is "the way we live our lives."

Islands, lakes, rivers, mountains and other topographical features were given names with much consideration to function, origin, practical use and in relation to significant stories in human history. It was not a custom to name places after oneself, as occurs in some other cultures. For example, Kennedy Lake is known as Ha-ooke-min to my people. Ha-ooke-min is a descriptive term. This name speaks of the bounty of fish and other resources that once filled the lake before the mass interruption of natural cycles. Upper-Kennedy River is named Winche, which also describes the resources that were found there. On the other hand, people are often named after places. This practice is a part of the traditional land titles law. Land titles were passed down through the generations from titleholder to successor and so on with the passing of names.

Wearing the name of a place comes either as an inheritance within a sub-family group, or earned through expert management of specific land and water holdings. Either way, it carries considerable responsibility. A name-bearer's mismanagement of a holding would not be tolerated by the laws of our people. Therefore there is an emphasis on a careful education process for those wearing place names; ownership of the land implies a responsibility to protect it and use its resources prudently, not a right to exploit it or otherwise misuse it.

LOGGING IN CLAYOQUOT SOUND: HUMANIZING THE INDUSTRY

Cedar has always been a major staple in Tla-o-qui-aht way of life. Harvesting methods were not mechanized; instead, approaches were more labour intensive and restricted to meeting the needs of the community. Certainly things have changed dramatically over the past fifty to seventy-five years; however, I believe that we now have the opportunity to apply our traditional principles to modern tree-harvesting practices in Clayoquot Sound under the following guidelines:

- that methods of tree harvesting slow down with a focus on creating more local jobs that encourage healthier lifestyles;
- that all trees harvested in Clayoquot Sound stay in Clayoquot Sound with the exception of value-added wood products that are attributable to local job creation initiatives;
- that each tree harvested be replaced through local job creation initiatives that integrate cultural values of forest stewardship.

The underlying principle of these three guidelines is a focus on humans rather than on volume or profit. That is not to say that this approach excludes profitability — it can foster companies that are quite profitable. However, the benefit would stay local, business owners could make more money using less wood, and there would be more jobs available within our territory.

LOOKING FORWARD: TRIBAL PARKS AND RESTORATIVE JUSTICE

Tla-o-qui-aht are at a unique point in our history. There is a parallel between the disappearance of old growth forests, the scarcity of our salmon runs and the changes in our traditional ways of life and loss of language. In the past hundred years, we have come back from the brink of extinction and we're stronger for it. In moving forward, there are two main challenges facing Tla-o-qui-aht that will shape our approach to the next part of our history. These are the following:

- healing community conflict: resolving the inter-generational effects of the residential school experience;
- healing our traditional lands: restoring the natural balance of creation and consumption in our lives.

Tla-o-qui-aht are taking the responsibility to address both of these issues through program development under our Tribal Parks Establishment Project.

THE TRIBAL PARKS ESTABLISHMENT PROJECT

In this project we are building on all of the work that was done by our elders and community leaders during the Meares Island court case and during subsequent treaty negotiations. We are doing so by integrating all of the input regarding our cultural teachings and practices, which was collected into programs such as the Nee-waas-nish Nisma Restorative Justice Program.

THE NEE-WAAS-NISH NISMA RESTORATIVE JUSTICE PROGRAM

Restorative justice represents a fundamentally different approach to crime and conflict in communities. Rather than punishing offenders, proponents look to address the underlying social problems that lead to crime and conflict. The restorative justice approach can be applied to all stages of the conventional justice system from enforcement to correction and reintegration. In our case, we focus on crime prevention through cultural revitalization, as an adoption of the more conventional crime prevention through social development.

TLA-O-QUI-AHT TRIBAL PARKS

The primary goal of the Establishment Project is to develop all the soft infrastructure required to give life to a formal Tla-o-qui-aht

Tribal Parks organization. As mentioned, much of the community consultation has already been undertaken in previous processes, and additional consultation is presently being undertaken within Tla-o-qui-aht and also throughout the broader population within our traditional territory. The organization's philosophy of social and environmental restorative justice seeks to address the underlying issues that lead to conflict in our communities through cultural revitalization, and by strengthening the connections that we have with the land.

Nee-waas-nish Nisma is a term for joy in heritage. It is about building on our strengths and taking responsibility for our feelings and our future. It's also about taking back our rightful place as stewards of the land. Most importantly, program activities cultivate a self-image that is consistent with the teaching that we are a link between the ancestors and future generations.

This world view is rooted in our responsibility to care take of and pass down the vital staples of our way of life here in Clayoquot Sound: cedar and salmon. The Tribal Parks Establishment Project team is actively engaging with the non-aboriginal residents of Clayoquot Sound as well. This aspect of the organization reflects one of the founding principles of our culture: we are all Quu-us.

REFERENCES

Arcas Associates, Heritage Resource Consultants. (1986). *Patterns of Settlement of the Ahousaht and Clayoquot Band Vancouver: Book One to Five.* Port Moody/Kamloops, April 1986.

Gough, B.M. (1971). *The Royal Navy and the Northwest Coast of North America 1810–1914: A Study of British Maritime Ascendancy.* Vancouver: University of British Columbia Press.

Jewitt, J. (1815/2000). *White Slaves of Maquinna: John R. Jewitt's Narrative of Capture and Confinement at Nootka.* Surrey, B.C.: Heritage House Publishing.

Mathes, V.S. (1979). "Wickaninnish: A Clayoquot Chief as Recorded by Early Travellers." *The Pacific Northwest Quarterly,* July 1979, 70.

McMillan, A.D. (1999/2000). *Since the Time of the Transformers: The Ancient Heritage of the Nuu-Chah-Nulth, Ditidaht, and Makah.* Vancouver: University of British Columbia Press.

NOTES

[1] Tla-o-qui-aht oral history (the date was determined by calculating the generations through which names were passed).

[2] See "Patterns of Settlement of the Ahousaht and Clayoquot Bands," Book 1 of 5, p. 65. Produced as part of the process to place a court injunction on Meares Island.

[3] See *Since the Time of the Transformers: The Ancient Heritage of the Nuu-chah-nulth, Ditidaht, and Makah,* pp. 186–187.

[4] See *The Royal Navy and the Northwest Coast of North America 1810–1914,* page 11. For more general information, see "Chapter 1: The Contest for the Columbia Country, 1810–1818," pp. 8–29.

[5] See Gough, Chapter 1; Mathes, p. 118.

[6] Tla-o-qui-aht oral history.

[7] See *The Royal Navy and the Northwest Coast of North America 1810–1914: A Study of British Maritime Ascendancy,* pp. 9–11.

[8] Tla-o-qui-aht oral history; Mathes, p. 19. For more information see J. Jewitt's *White Slaves of Maquinna.*

Excerpts from

Clayoquot: The Sound of My Heart

- BETTY SHIVER KRAWCZYK -

Cougars are normally very shy creatures, and in the past their reputation has certainly been that they will go out of their way to avoid humans. But something pitiful has happened to them of late. Some of the old homesteaders say it is the clear-cutting on the island that is making the cougars crazy.

Clear-cutting is the logging method used by the giant multinational logging companies like MacMillan Bloedel and Interfor. In this method every living tree is cut down in a given area. The unwanted trees and brush left behind are often burned, leaving great open spaces that are foreign to the hunting habits of the cougar. Pack-hunting wolves, however, thrive in this new, unexpected windfall of open spaces.

Reprinted by permission of the author. Originally published by Orca Books, 1996.

The wolf and the cougar both feed on the small black-tailed deer that used to abound on Vancouver Island. The wolf thinks clear-cutting is great, because s/he catches prey by chasing deer into the open and simply running them down. The cougar, however, is a cat, and like all cats, hides and stalks and pounces. As there is nowhere to hide in a clear-cut, no tall trees from which to leap and pounce, the cougar, like the North American Free Trade Agreement, is looking for new, mostly non-existent markets. And so it is the odd cougar, the young ones who are newly on their own and not-yet-great hunters, or very old ones who know how to do it but haven't the physical strength to carry through, who turn their attention in the direction of isolated homesteads and villages. They prefer small pets that can be taken with a minimum of fuss.

I have no pets and the only time I actually bend all the way over, at least under a dense tree canopy, is when I have to fill the water jugs from the stream. As I will have to do shortly. I pick up the water jugs and with one last glance at the safety of the A-frame, start off down the pathway to the stream.

I love the A-frame. It's a home-made house. My son Mike built it all by himself, except for the actual erecting of the frames which required a sort of barn-raising beer party. The A-frame is perched on a rocky bluff and when the tide is in I have the wonderful sensation of living in a houseboat, surrounded by water. I used to dream of living in a houseboat. In a free-floating houseboat one is totally out of reach of one's creditors in the immediate sense. At least they can't drop by or call you on the telephone. And if one doesn't like the neighbourhood, one can jolly well mosey on around to another cove. Houseboat-living precludes being arbitrarily summoned by one's children or grandchildren for this or that little emergency. The dear ones must survive on their own until grandma wants to emerge. Still, I have most of those advantages now, perched as I am atop my rocky bluff. And when the tide is out I have the additional sensation of being suspended in space. The structure is just one huge room with sleeping lofts upstairs but there are lots of windows and skylights, and wherever I stand or sit inside the house I can see through to the outdoors. Which makes me very happy.

I don't know what it is with this wilderness gene. If I had known from the beginning what kind of life makes me happy, I could have saved myself a lot of misery. I was raised in the country, in East Baton Rouge Parish, in the state of Louisiana. I loved the country, but the aim seemed to be, on everybody's part, both black and white, to get out of it as fast as possible. It's hard to hang onto something that is so downgraded by everybody else. In fact, my brother and I couldn't beat the country dust off our shoes fast enough. Downtown. Just show us the way. The funny thing is, both my brother and I have since spent a lot of years trying to get back to the heart of nature.

But one really can't go home again. My brother has retired halfway up a mountainside north of Phoenix, Arizona, which is a different face of nature, one a long way from the swamps of Louisiana. And while Clayoquot Sound evokes some childhood memories for me — the heavy rainfall is sweetly familiar, as is the fresh, seaborne air — still, most is strange. It's like meeting a new man, the strongest attraction is centred on the one with the most challenging blend of the strange and the familiar. Only my fascination with the Sound hasn't worn off. At least not yet. I have lived in some impressive places in my life, as well as some that were only a step or two above the hovel category, but none has filled me with such joy as this place. If only I didn't have to go get the damned water.

Before I became intimate with Cypress Bay, which our little cove claims to be part of, I knew little of pebble beaches. A beach was supposed to have white sand and at least be good for wading, if not swimming. Our little beach cares for none of these affectations. This beach is a sort of marine Calcutta, swarming with life, each organism trying to survive, often under great odds. There seems to be an abundance of squabbling and death-dealing altercation between the different species, but there are also some manifestations of tolerance, if not downright cooperation. I put down the water jugs and pick up an oyster shell.

The shell has several rows of barnacle townhouses, all occupied. A single oyster shell can carry dozens of barnacle townhouses, large and small, over its surface. As these particular tenants sense that their entire planet is in danger, they decide it's time to hunker down and play dead. What curious creatures they are. Do they imagine their planet, their world of the oyster shell, to be flat? It is, rather. It certainly isn't round. Perhaps, like the barnacles, we believe our world to be flat, but unlike the barnacles, are convinced that it is round, or not quite round, by our scientific elite. The barnacles presumably have no scientific elite to confuse their perception of their own particular universe. Then again, perhaps our universe, scientists and all, is really no bigger than the back of the monstrous sea turtle that the ancient Chinese sages imagined the world rested upon. Maybe the ancient Chinese were right in principle, they just didn't think in universal terms. After all, does anyone know for sure what lies outside the universe of our universes?

I carefully place the shell back in its original position. Is that a small collective sigh of relief I hear? The seaweed is full of tiny hitchhikers and the stationary rocks are smothered with scallops and oysters and mussels. I am always on the lookout for interesting shells because I make little craft things. I don't have the particular gene that makes one good at that sort of thing, but I don't believe in letting a lack of innate ability get in the way of expressing myself creatively. However, it seems almost impossible to find a large shell that isn't housing a bunch of squatters of one sort or another and this is definitely hindering my craft-making.

As I pick up the water jugs to resume my journey, I glance across the cove. The bears have come out to play. More precisely the bears have come out to eat. Small black bears, two of them. Teenagers, I think. I've seen them at least a half dozen times in the last few weeks, but always through binoculars from the safety of my front deck. I think their mother has kicked them out. They are not so close to me as to be threatening in any way. I stand motionless, watching.

I must be downwind of the bears, because they give no indication that they are aware of an audience. They are totally focused on the task at hand, which is to try to turn over one of the rocks under the bluff by the stream. This is the largest of the streams and I can hear the sound of its waterfall from my front deck if the birds aren't making too much racket.

The bear mother has taught her children, before she decided these big kids were old enough to fend for themselves, that the outgoing tide left little presents underneath the rocks. Tiny crabs and other dainties. But the bears must wrestle the rocks for their treats. When one particularly large rock won't yield to the strivings of one, the twin comes over and offers assistance. The rock is stubborn. The bears have each taken turn and turnabout before the rock finally gives up and rolls over. The bears pounce on the tiny, luckless crabs now pitiably exposed, and after a couple of moments of intense feeding, amble down to another cluster of rocks. But the wind has shifted. The hateful stench of humans has evidently reached one of the twins. S/he rears up on hind legs and looks in my direction.

I doubt if the bear can see me with its poor eyesight. But this one is definitely spooked. S/he suddenly turns and bolts for the rocky bluff that fronts the towering cedar trees. The other follows, lickety-split. They aren't taking any chances.

The seabirds don't give a damn. They are fishing too, and I almost have to kick them out of my way, they are so intent at table. They remind me of my kids, when they were all at home and mostly made of mouths and stomachs. But I like them, raucous bunch that they are. Even those greedy-guts, the seagulls.

But the loon is my favourite. I love its romantic, mysterious cry. The diving ducks are in some rare category of marine maladaptation. Just from watching, it seems to me that ducks, like the loon and the canvasback, need an inordinate length of runway in order to get airborne. If they are startled while in the water and try to make a quick exit, they splash and flounder and half-drown themselves. The mallards and pintails don't dive for their food, they just tip upside down in the water and show their cute little behinds. But they're much better designed for emergency takeoffs. The haughty blue herons look great in the air, sweeping low on their wide, wonderful wings, but, when they land and go beachcombing, they walk with the peculiar hesitant gait of the stork. There are two families of blue herons living somewhere around our cove and they bring their youngsters here for fishing lessons. They all have better luck catching fish than I do.

The bald eagle likes our neighbourhood, too. Everybody likes to watch the eagle. Monarch of the mountain. The power and grace of the eagle's flight and strike is awesome. However, the bird has a wimpy cry. A kind of whistling sound, more suitable for a smaller bird. One of nature's little trade-offs, I guess. A kind of reining in of the high flyers. Mother Nature fixed it so the eagle wouldn't get a swelled head in both the looks and voice department, otherwise the bird might think itself the favourite of the skies, like the human male thinks he's the favourite of the earth. I mustn't tarry any longer. Duty is calling.

I'm sure there are no clear-cuts in heaven. There couldn't be, they're too damned ugly. The mountains look like they've been napalmed. But maybe I just don't understand this clear-cutting practice. The man representing the logging company that sold us our ten acres said the mountains had been replanted and would soon be sprouting a new forest. Well, I knew then that there must be a difference between an old growth forest and a tree farm, but at the time I didn't understand to what degree. And as I gaze daily upon these stripped mountains, I am beginning to wonder if the logging companies, after all, know what in tarnation they are doing.

At the stream I blow several ear-splitting blasts from the whistle around my neck. It's one of those high-powered whistles, guaranteed to intimidate man and beast alike. The birds rise in one body like a shot — at least, the ones that are designed to do so. The others flail across the water like so many faulty little motorized scooters, and I can hear the sudden scurrying sounds of the small panic-stricken land animals trying to vacate the area. Quickly, while every living thing about is still stunned, I stoop down and begin to fill the jugs with the clear, gurgling water.

On my second trip to the stream I notice that the wild rose bushes on the opposite bank are weighted down with hundreds, no, thousands of small, heavenly scented blossoms. As there are obviously no cougars out today, I make a third trip for roses and asparagus.

Sea asparagus is a long-stemmed, cranky-looking plant that resembles chicken feet but is tender and tasty when lightly stir-fried. It grows wild in thick clumps at the high tide line. As I head home laden with flower and sea plants, an eagle swoops low, makes a complete circuit of the cove and heads away over the mountains. What do the eagles think of the clear-cuts, I wonder? The clear-cuts must seem very strange to them, to all of the animals, in fact. After all, a good hunk of their world has disappeared.

But I have wild roses, sea asparagus and pure, sweet water and I won't let those poor scalped mountains intrude upon my enjoyment of an incomparable summer day in July in the Clayoquot Sound.

At the time, of course, I have no inkling that those same wounded, bleeding mountains — on another incomparable summer day in July several years later — will throw me into a conflict so passionate, unrelenting and uncompromising that I, a perfectly respectable if somewhat eccentric grandmother, will wind up behind prison bars, condemned as an enemy of the state.

[EDS. In the next section, Betty returns to Cypress Bay after more than four and a half months in jail for blocking a logging road in Clayoquot Sound.]

I don't know what the result of the protests and trials will be, but hopefully a new way of looking at nature will emerge, a new regard for the natural world that will incorporate within itself the awareness of the laws of nature, a recognition of the fact that we cannot, as a species, simply destroy the lifelines that sustain us, and that natural law and judicial law will somehow meld and become a new thing.

My body aches from being tossed hither and yon these last months, and my spirit craves respite from the confusion. The cove holds out her leafy arms and enfolds me as the tide gently washes me ashore. I am sorry, I whisper as I embark and start up the winding stairs to the house. I'm sorry. I tried. I tried as hard as I could.

The cove answers, and the mountains. I hear the whispering, the gentle murmurings as the night settles down about. What is being said? I don't know. In the days that follow, I listen intently to the language of the cove in the silence of the evenings, but I am not being spoken to, perhaps spoken about, but not spoken to. The perfect unity I had felt at times with this place, that perfect peace, escapes me now. I busy myself with the physicality of my days, trying not to think too much. There is firewood and water to be hauled, clams to be dug, fish to be coaxed onto the line. There are books to be read, a spring garden to be planned. At Christmas, the family gathers in Vancouver at Barbara's place. Immediately after, I head back to Cypress Bay. As the winter flirts with spring and then succumbs completely, I know I am still not right with my cove. Even after the bears stumble out of their winter lairs and amble about looking for food, the bird population returns in full force, the giant slugs are lurking underneath the planter boxes waiting for the first sign of garden activity to begin, and the salmonberry bushes are just starting to bloom, I still have the peculiar sensation that the full face of the cove has turned from me.

I think the cove is turning her face because I am no longer very optimistic about saving the Sound from clear-cutting. The cove senses my despair. She does not share the innermost secrets of her soul with summertime soldiers. But how can I be other than pessimistic when I see that the logging of the ancient ones is continuing non-stop? What can one do, I demand of the mountains behind the cove, after one has done the very best that one can? Occasionally I yell at the cedars down by the creek and ask what *they* would have me do? I'm only one person, for Pete's sake, and an old lady at that. I'm tired and I just want to retire and be quiet. Can't you understand that? But the trees are dumb and so are the rocks and the streams and the mountains, and even the Steller's jays, who are the biggest beggars and chatterboxes nature has ever designed, seem to be avoiding me.

How stupid and selfish of me to despair, even for a moment, for I am a product of nature, too, and my voice is also nature's voice. Forgive me, mother of the universe, for slacking off and losing faith, I will use nature's voice within me with all my heart and soul, oh, yes, I will, for as long as I shall live.

A Groundswell is a Wave

- VALERIE LANGER -

And forever before me gleams,
The shining city of song,
In the beautiful land of dreams.
But when I would enter the gate
Of that golden atmosphere,
It is gone, and I wonder and wait
For the vision to reappear.

— "Fata Morgana" by Henry Wadsworth Longfellow (1873)

I came down the mountain, dust and stones in my shoes, Andy and I railing about the damage. His English was dressed in Scottish brogue. We'd gone to see where giant trees had been felled earlier in the year. Across Clayoquot Arm, ancient forests rose from the mountainside that climbed from shore to ridge on a steep angle. We stood on the ugly side, cut through by logging roads, mangy in its cover. Straining our necks back, we looked up the slope. Emily Carr's "screamers" — sharp daggers of wood where trees broke off at the end of their fall — sat jagged across the massive cedar stumps. Great shards of wood lay all akimbo from the logging. No mist rose like smoke wisps from the clear-cuts. It was an unusually hot and still morning.

When the tall ships sailed the tartan clans off to the New World, the boreal forests of Scots pine and oak had already been turned to moors. The songbirds had given way to grouse flushed out to be shot, and to golfers. The forest had been hushed. Nine hundred years ago a hunter saw the last wild bear in Britain. The wolf evaded the bow and the gun for another 600 years, skulking around the edges of survival, and losing. The deer and the heather flourish in Scotland's landscape that we now take for natural.

Here in Clayoquot Arm the forest was being hushed. We walked down the logging road back to the old Volvo that served as our off-road vehicle. "We've got to reforest Scotland with its native species and reintroduce the wolf to keep the deer in check," Andy said. "The country has been so impoverished mentally and environmentally by its history of the removals, people can't remember what they used to be and what they used to have. As a result, many Scots are opposed to the idea of bringing back the forest. They want to keep the moors now." We stopped to look at a mound of bear scat purpled by berries. "It looks like jam," he said, then, staring at the evidence of the mammal predator and the succession of wild fruits, he concluded his thought. "People get used to the poverty of nature that surrounds them."

Far below, brilliant light scattered across the surface of Clayoquot Arm, a part of Kennedy Lake. Farther out, the ocean had a molten glow. I couldn't tell the tide was swelling.

The return to Tofino brought us across the Kennedy River Bridge again. I stopped so Andy could look over the location of the mass uprising of 1993. There was no evidence of the groundswell against logging; no evidence, this late summer morning by the lake, that so many of us had been arrested here two years before. It was just a quiet place with a big history.

From around the bend a car emerged and drove up to the west bank of the river. A lanky man pulled himself out and began to adjust his camera. "Hello," we said, because we were only the three of us, a long way down a logging road on a quiet day in Clayoquot Sound. He looked at me and said in a German accent, "I know you."

I looked at him more carefully to see if I could conjure some recognition. He was tall and brown haired and utterly a stranger to me.

"I have seen you make a presentation in Osnabrük the year before," he said, his English heavily inflected. "I am here now to photograph this famous place."

Some Europeans come to exalt and others have come to exploit the biological wealth for which British Columbia is so famous. About fifty years ago a European immigrant to British Columbia invented an efficient way to simplify the complex forest. He had probably reeled at the riot of giant biodiversity in the moist forest community of the coast. It was the antithesis of the open order of the European second- and third-growth woods he knew. Wanting to make himself and all the other immigrants feel more at home in their new surroundings, and with profit from the resource in his sights, he projected a simplified forest, in rows, for the second cut. The forest was divided into parcels, each assigned a generous amount to be logged every year. This system was to be managed neatly for a humanly conceivable rotation of eighty to 100 years. It made great sense to the industrial European mind.

Ten thousand years ago, the retreating glaciers were shaping Canada's temperate rainforests. Across the ocean those millennia ago, tribes from the south began to populate the caves and valleys of the European continent, foraging for nuts and meat in the developing woods. At the same time, under similar circumstances the Nuu-chah-nulth culture was evolving with the changing forest and sea of Clayoquot. The salmon arrived with the cedar, which grew into massive living sculptures. Hemlock grew furtively in the shadow of their charismatic cousins. After ten millennia the patterns, established, delivered a yearly run of salmon up the rivers, and the forest cycled in increments of a thousand years or more.

In Andy's homeland, the Scots pine forests that had developed after the melting of the ice sheets were cleared centuries ago. Fifty years earlier, Canadian foresters began a project to replant large areas of the former Scottish boreal forest with imported Sitka spruce. Sitka belongs in B.C., growing crooked in the salt spray of the shoreline and towering up from the bottoms of river valleys. I had met Andy, an ecological historian, when he organized a temperate rainforests slide show tour in Scotland for a colleague and me. Near Aberdeen he took us to hike in one of the few remaining indigenous pine forests. After twenty minutes of walking through heather, under the mix of conifers and deciduous trees, we suddenly walked out onto the wide-open moors. After a few further steps we turned to look back at the edge, straight as an arrow, that defined the end of the natural forest. That was it.

Over several days, I had taken Andy for a number of hikes in Clayoquot Sound. In the dark woods, bars of shadows and light slanted through the canopy. The trees rose out of the mass of berry bushes, ferns and devil's club. Many of them stood rotting on their roots, slowly becoming the next generation of soil. What seemed like death was generating life, and life was organized into a vast complex, deep, wide and tall. Invisible threads of fungi and uncountable insects worked the system, their presence visually overwhelmed by the grandeur of the vegetation. The ravens made an eerie call that sounded like stones dropping down a deep well.

In contrast to Aberdeen's remnant pine woods, the Clayoquot rainforest is a cathedral of timber stretching from the seashore to the alpine rock. And invasive concepts from the Old World have arrived. The gun, the chainsaw and the idea of money exercise themselves across whole landscapes. The most beautiful expression of life is reduced to a fibre commodity and simplified robotically to plantations, if it grows at all in the weakened soils. Evolution is truncated. I rage at the thought. Here, at least, the forests still lie in folds over the mountains and along the inlets. They have skirted extinction for the time, as Britain's wolves once did.

Given the time this forest has been in Clayoquot Sound, my status here as a local seems absurd. Two days, two weeks or twenty years are insignificant in evolutionary time. But my life did change from the moment the police took the first arrestee from here two years before.

I had stood day after day on this same spot at the bridge. I had introduced the Belgian member of the European parliament to Canadian MP Svend Robinson here just before Svend was arrested for civil disobedience. The now (in)famous Betty Krawczyk and three other grandmothers had linked arms with him and several other young men and women. All had been strangers to each other, and all launched themselves and Clayoquot into the international sphere with the simple words, "No, I will not move." Below the sounds of dissent, the press cameras and tape recorders had clicked, rolled and whirred, day after day, as hundreds and hundreds stopped the logging trucks, were arrested, tried and jailed.

For three months I would watch the sun rise at the bridge every morning with hundreds of others. I would barely sleep for six months. The landscape became politicized. People came pouring in from all over the island, the country and the world. The bridge in Clayoquot Sound was where the tide began to turn on industrial logging.

"You know," the German man said, "there is a tsunami alert for today." We focused on him, suddenly the most earnest man I had ever met.

"What?" Andy said.

The blood pulled quickly into my core leaving me cold for an instant. "A tsunami!" I said. "When?" My parents were at my house on the other side of the low spot at Long Beach.

"From the earthquake in Kobei City, Japan, last night, they are saying a tsunami comes here this afternoon," he said. "I am leaving to the other side of the mountains after here. Many people are leaving now."

I stood for a moment on the Kennedy River Bridge in the stillness of the day with a German and a Scot, our minds turning to the potential of a tidal wave.

Andy and I had been climbing around clear-cuts since early that morning. Now there was urgency to return home and figure out the best high ground. My parents were there. I couldn't consider just heading high up the mountain. I urgently needed to connect with them to make sure we all had a good plan for whatever nature was bringing us that afternoon.

Andy was to fly out the next day, back to the old country. As we

West coast of Vancouver Island. (PHOTO: JEN PUKONEN)

A beach sunset near Tofino. (PHOTO: JEN PUKONEN)

drove the road by the oceanside our conversation was disjointed, interrupted by sudden questions and macabre laughter. Knowing only the briefest details of the incoming wave, we had the nervous luxury of time to consider a plan of action. We rambled into plans for re-establishing the forests of Scotland, saving the forests of Clayoquot Sound along with figuring out how to get Andy and my parents onto planes by tomorrow afternoon. How to save ourselves from drowning figured as well, but we knew little about what might be happening. As we came over the rise before Long Beach, we looked out over the surf to see if anything seemed different. The waves were rolling in their gorgeous and ageless pattern. Andy stared out with his intense deep-set eyes. "Bloody beautiful, isn't it?" he said, and then we descended below the canopy onto the straight low stretch of road that parallels the beach.

Just then two trucks pulling whitewater rafts passed us in succession, their urgency apparently greater than ours. The young men in the trucks had their windows rolled down. I wondered if they knew about the tsunami but couldn't really shout out to vehicles passing at such speed. They may have been rescue boats, but they looked like whitewater rafts, we thought. The road cut through the forests, crossing every salmon stream that flows through Pacific Rim National Park. We crossed them all on a highway that could soon be washed away.

When we finally pulled into the driveway, my mother and father were relieved to see us. They were agitated. They were booked to fly to Panama the next evening. That was the only fixed decision of a two-month trip they were embarking on. The start date was set but there was a problem. My mother's new passport had not been delivered. It was to have been couriered days before and now, with a tsunami looming, they were facing a narrow set of choices. They would have to leave before 2 P.M. when the highway would be shut down to prevent anyone from being washed away on the low spot at Long Beach. Leaving for Vancouver before the wave would get them closer to the airport, but would not get them onto the plane without passports.

"Pack up and go with Andy," I said.

My mother was distraught. She was in an impossible bind, exacerbated by my refusal to leave with them. Time was closing in, and a decision had to be made. We argued back and forth whether I should come with them. My father and I felt that the house was on high enough ground, according to information he had picked up from the radio and talks with people in town. My mother has always taken risks and allowed us the same latitude, but this time she preferred that we all to be together. It was not about "calculations."

We gauged how high above sea level the house was. In my mind I had an image of a massive wave tearing down the trees half a kilometre away and eight metres lower down at Mackenzie Beach. The trees would intercept the tall water wall and I would be safe. There was also Barr's Mountain and the top of Industrial Way, all accessible high points. In town, people were buying beer to take atop the Maquinna Hotel for the best seats from which to view nature's best show.

My mother packed up the last of the bags, and we loaded them into the car. She was weeping. Their visit was ending in a tumult of decisions, apprehension and separation. Andy was ready to go, also leaving a day early. I needed to prepare, to fill jars with water and buy some bread in case my home became an island without a drawbridge.

"You need to go," I urged.

"This is ridiculous," Mom said. Dad began picking berries beside the fence. Andy was shifting awkwardly. Then a vehicle turned into the driveway. The courier, a soft-shaped man with an easy smile under his dark moustache, hopped out of his yellow truck. Blithely he walked into the emotional mayhem of the yard.

"And which of you is Shirley Langer?" he said cheerily.

My mother stepped leadenly towards him. "My passport," she said, simultaneously deflated and relieved. He handed her the small package and turned the clipboard towards her.

"Sign here for your passport out of here," he chuckled. As soon as she signed, he bid us goodbye and efficiently continued on his way, turning right into town to deliver goods.

They had to leave immediately. The three of them, my parents and my friend, climbed into the car after hasty embraces. My mother was weeping again. If it weren't for the fact that there was nowhere else I would rather be, I would have gone too. I stood waving as the car turned left onto the road for the long drive out.

The bike ride to town was beautiful, save for the traffic. Another truck passed me towing yet another big river raft. There were cars driving out of town to make the highway before the closure and cars driving in to get provisions. Across the inlet, Meares Island sat like a Buddha. The air was so clear that each tree was distinct on its mountainside, even from a distance. The harbour was calm, without portent of any disturbance.

A fourth truck with a raft passed me, then slowed to a quick stop. A young guy stuck his head out the window and flagged me. "Where's the boat ramp in town?" he asked.

"Back at Fourth Street. Down the big hill. There's a tsunami warning, you know?"

"Yeah," he grinned. "We're going to ride it."

I rode down Second Street to take another look at the water before going to buy bread. A poster taped to a telephone pole caught my attention. In bold capital letters it said TSUNAMI ALERT. Below was written the ETA, notification of the impending highway closure and some other information about high ground. I cycled to the building environmentalists had dubbed the MacMillan Bloedel "Disinformation" Centre. There is a good view of the harbour from there. A logging crew boat was tied up to its private dock. I noticed one of the rafts that passed us on the road speeding out to the ocean.

From the old logging company office, the view out to Meares and Catface Mountain is spectacular. The sight of the greatest plant biomass per hectare of any ecosystem on Earth is impressive each time. My clear sensation of wonder was invaded by the thought that anyone would consider razing the whole scene. Even in this little paradise town we have wrangled for years over which value is the greatest: the biological or the economic.

Money grows on trees, but it is a brief fruit. It has taken a little over two generations to undermine most of the ancient forest on Vancouver Island. An embarrassment of riches has been generated. The railroads transported logs south to the US, sometimes only one or two quarter-lengths of a tree fitting on each rail car. Great tanker ships, their holds filled with 600-year-old red and yellow cedar, crossed the ocean to Japan. I have witnessed the self-loading leviathans wait in the bay for trucks to bring the diminishing bounty down from ever more marginal tracts.

Almost every town and city has been geared to moving the richness away from the hillsides. The air withers. Stones and earth slip down the valleys and wash into the ocean. Birds seek new nests in places already over-occupied by their species. The edges of forests dominate instead of the protected centre. Radical changes lay waste to the landscape and people see castles instead. *Fata Morgana.*

A revolt was only a matter of time away.

Looking out at the water gave no evidence of what might happen. It was an exercise to see if I could see change coming. Half a day earlier an earthquake shook Kobei City at 7.2 on the Richter scale. The solid ground and the ocean floor shook with ferocity, surprising and killing people while we slept over here. Below the surface a sine wave of energy began to move eastward, collecting force through the ocean. It headed towards Hawaii. The Coast Guard in Tofino was alerted to post warning signs in all public places. The town moved slowly in preparation. We discussed where to be. We bought bread and shopped for a few provisions. We considered Radar Hill, the roof of the Maquinna Hotel and our homes as vantages or bunkers.

A tsunami is not an ordinary wave; it's a wave of energy producing a flood. It builds into a wall of water, emerging quickly as the flood is pushed up the shallows near shore. The concentration of water is recruited from the edges of the ocean giving the appearance that the shore has emptied. And then it arrives, shockingly reoccupying the coast. In politics such energy is called a groundswell. Usually political groundswells are composed of relatively few people possessing an enormous force. They always seem to surprise people, not least those who put the energy into building them. Like most of the ocean during a tsunami, the majority of the population is uninvolved. There are always hucksters, unengaged or antipathetic to the movement, selling memorabilia and snacks to the crowds. And there are those who join in the experience of the energy high. The bulk slip back into the muddy waters once the work of rebuilding begins, slower, more tedious and completely changed in its energy after the presentation of the wave.

I watched another raft pass by: more thrill-seekers on their way out to sea. The eagle that had been perched in the Eik cedar was wheeling around over Dead Man Island and the cormorants' rock.

Outside the Co-op a group of people were talking about the last ETA update. The wave was only a foot high when it passed Hawaii. The mood turned suddenly celebratory. Danger had been thwarted. The shakeup had not been big enough or close enough to reach us this time.

I swung up the hill to the bake shop but couldn't get by one of the Coast Guard, who was taping something across the alert posting. He finished quickly and left with the sheaf of updates to be stuck across the other alerts. In thick uppercase type, taped on an angle across the sign I read "THE TSUNAMI HAS BEEN CANCELLED."

Later that afternoon a small flood of water washed up Chesterman Beach, almost reaching the driftwood. An unscheduled high tide. The sunset turned the trees purple-orange. The heat of the day dissipated and the ordinary returned, save for an undercurrent.

Thirty Years in Hesquiat

- DIANNE IGNACE -

Dave first brought me to Hesquiat in February 1975, supposedly for the weekend. We were not married yet. He wanted to gauge my reaction to the place, as he couldn't ask me to marry him if I didn't approve of his indigenous territory, which is quite remote and not for everyone. Hesquiat is on a peninsula that extends out to the open ocean. It is accessible only by boat or air, and there are no other communities within a reasonable distance. Fortunately, I fell in love with it instantly.

Quite unexpectedly we were stranded for ten days on that first visit. After dropping us off, Dave's cousin had an accident onboard his troller at the fishing grounds. He cut his forearm with a dressing knife and left the area immediately for the hospital back in Tofino. We waited for the troller to come back for us, but as Dave's antique crank phone was dead we had no way to contact the outside world — not even a hand-held V.H.F. (Very High Frequency) radio. My brother in Campbell River thought I had disappeared and was worried sick. In those days, at that time of year, there weren't many boats. A family in Hot Springs Cove knew we were out at Hesquiat, and they sent a speedboat, which belonged to another of Dave's cousins. Until then we had to wait, without knowing what had happened.

Dave and I quickly came to know each other very well. In those ten days, I received a first-hand survival lesson, as we had brought only enough food for *two* days. Dave's many talents for running his cabin were revealed: he had to start the gravity-fed water running, get the generator operating smoothly, gather firewood, harvest seafood, hunt ducks and show me how to cook food I'd never seen before. I was in a constant state of amazement.

FINDING HOME

Our next visit to Hesquiat was in the more pleasant month of June, but Dave had to leave to go fishing far away, towards Kyuquot, and I was left alone. I wanted to sew my wedding dress, but panic was setting in because there was only a week left before the wedding. There was still no plan for cake or flowers, and I couldn't call my mother or brother to check on their travel preparations. Luckily, Stu and Sharon from Estevan Lighthouse came for a visit. Sharon had a sewing machine and invited me to Estevan for the night to use it. That was one of my few rides by truck on the board road to the lighthouse (later in 1976 the government closed the road). Once there, I was also able to make the calls needed to arrange all the details. Finally, when Dave returned, he arranged a ride out to Tofino for us so that we could look for a preacher! Nothing like hurried preparations.

Dave and I were married June 30, 1975, on Chesterman Beach in Tofino. The wedding was at low tide. The preacher stood in a little rock alcove facing our guests, who sat along rock formations, while elders sat on chairs on the sand. To my family, it was an unusual setting. My mother's father had travelled for his first and only time to Vancouver Island for the occasion. He was amazed at the winding highway, which he could describe only in German.

During the reception Dave's father, Grandpa Hesquiat, offered to share his home with us. After a honeymoon to Meota, Saskatchewan, to show my husband where I came from, we moved in with the old man. Our first daughter Jody was born in September of 1976. In 1979, with Grandpa's permission, Kaesok Jill (Kaesy), our second daughter, was named after the great grandmother of all the Hesquiats.

The summer Jeff was born, Dave travelled from Tofino by canoe with Dulcimer Dave and some Ahousat locals who wanted my husband along to guide them through Hesquiat territories. It was a seal-hunting canoe, and Dulcimer Dave had a second larger canoe steered by *my* Dave. They called it the *Yashmakats.* He had a grand time until the July 1st weekend when I called him home with pregnancy problems. Jeff was born in August on Friday 13, 1982. We have celebrated every Friday the 13th since!

During the first fifteen years of our marriage, Dave went out commercial fishing as a deckhand on a troller. He would be gone for a week to ten days per trip. Sometimes the crew would go as far as Kyuquot, and I had no means of communicating with him. I was very glad to have Grandpa for company and, in the summer, Dave's sister Patricia and her daughter Janice. Sometimes other family members would come home too. Often there would be six to ten people for weeks; one summer there were twenty of us for over a month, crammed into one house and a few tents.

We've somewhat tamed our spot in the rainforest over the years. We hunt deer, geese, ducks and the odd seal, even bear. To add to our store-bought diet, we dig clams, pick oysters, pull crab, pick chitons and gooseneck barnacles regularly. Octopus and sea cucumbers, abalone and scallops are now a rare treat, as ocean stocks are dwindling. We still catch many types of fish, especially coho and ling cod. Halibut, red snapper and tuna often grace our table as well. In the beginning I had to learn how to bake bread and become efficient at inventing recipes, as ingredients in many recipes weren't always available. I also had to learn enough about carpentry, electronics, mechanics, plumbing and chopping wood to be the local handywoman if Dave was away working. I soon taught all

I could to the children so that no one had to work alone. Now Jeff is efficient at everything he's seen done and more — a regular jack-of-all-trades. He has built his own cabin.

CLOSE CALLS

There were a number of close calls over the years. When Jeff was two, and Dave was away, nearing the end of a fishing trip up Kyuquot way, I was in the bedroom reading to our three children when I had the urge to go to the kitchen to check on things. To my alarm, foot-long flames were leaping out of the stovepipe, which had broken off close to the ceiling. I grabbed a water jug from the gravity-fed sink and dumped the whole thing into the cookstove. The fire didn't go out. I told Jody to take Kaesy and Jeff outside; then I ran to the bathroom for more water and grabbed the V.H.F. microphone, calling "Mushguy! Mushguy! The chimney is on fire in Hesquiat!"

I ran out the back door and climbed onto the wash stand with a five-gallon bucket full of water. Somehow, I hauled it onto the roof, clambered up, lifted the bucket to the chimney and poured it into the pipe. I knew we were on the edge of the house burning down.

The fire hissed and sizzled and finally went out. I was still shaking and holding the children when Dave arrived, speeding into the channel in a punt. They had been towing it behind the troller off Boulder Point when my call went out, so he jumped in and raced the rest of the way. After such a close call, Dave stayed home for several weeks and missed a fishing trip or two.

Another time, several boats and about a dozen fishermen were here with gill nets spread out over the yards for repair before heading out to fish off Nootka Lighthouse for dog salmon. They were around all day and the yard was bustling busy. The next thing I knew all had departed, and I was alone with our three children again. Suddenly we heard running footsteps outside, so I scooped up Jeff and hustled to open the front door. There were five wolves face to face with our dogs, all standing on the porch having a stare-down! At the sight of me, our dogs started barking and all the animals moved. They began circling each other, growling, snarling and snapping. My dog Himy jumped at the biggest wolf, somehow biting its hind quarters and hanging on in a death grip. They started running in tandem. I became mobile.

I handed Jeff to Jody, slammed the front door and ran for the gun rack, where I grabbed the shotgun and some shells. I told the kids to sit down and went out the back porch loading the gun. Throwing open the top half of the porch door I watched the animals heading into the bush below the clothesline, Himy still clamped onto the wolf. I screamed at them, raised the gun to about fifteen feet above the ground and fired. I loved that dog and didn't want to hit him or any of the other dogs.

What I hit was the clothesline tree. Bark flew everywhere and the animals scattered out of sight into the bush. About twenty minutes later the dogs came back. Himy was missing a toenail and an inch and a half square patch of fur, skin, and flesh off his flank. Little Bear and Biscuit were bleeding on their legs, and Joey's dog Fawn had a ripped neck. The smallest, Fonce, was unhurt. Again I was shaking in my boots but much relieved that we were all OK.

I called Joe, our friend at the lighthouse, just to have someone to talk to about what had happened. Little did I know that he subsequently called Dave and told him the whole story over the air. Well, I was headlines on the coast for sure! The whole fishing fleet heard the story, and I am still teased about shooting the clothesline tree.

OUR GARDEN

I was four months pregnant with Jody when we completed our first garden. My experience with gardening was acquired as a child helping my mother with planting, weeding and harvesting. The soil here is soot-black and midden-rich. There is an average of three feet of topsoil covering most of Hesquiat, so it is very fertile. This is unusual for Clayoquot Sound as a whole, as temperate rainforests are known for their poor soil — except where the original human inhabitants have enriched it with centuries of shell piles. Hesquiats — my husband's ancestors — have lived here for at least four thousand years according to local history and archaeological findings. We have collected many artefacts, especially while gardening.

I now have sixteen gardens, several of which are more or less perennials, including two rhubarb and one potato garden. Cabbage, kale, onions, garlic, carrots and broccoli can be grown most of the year, especially with a little shelter from the rain. I have thirteen different types of plants from which to make tea, some for medicinal purposes but most to save on tea bags and provide variation in taste and diet. I always grow thyme, oregano, celery, chives and sage. In addition, I often have other herbs like summer savory and tarragon to cheer up our meals. We use seaweed for fertilizer, and we compost all organic wastes other than human. We have a septic tank and seepage tank for that.

HALLOWEEN IN HESQUIAT

Everyone on the coast seems to make a big deal out of Halloween. We used to go to town or Ahousat so the kids could go trick-or-treating. At least that's the excuse. Actually, I think Dave and I had more fun than the kids at Halloween. But on one occasion we were storm-bound at home with another couple, their daughter of Jody's age and our three kids. To make the best of it we devised a plan. We used salad bowls filled with cookies, raisins, marshmallows, nuts and a few other goodies we dug up. Then we dressed the children in extravagant costumes and explained to them that they should go knock on each door in the house and see what would happen. Meanwhile each costumed adult hid in a separate room and waited. As there were only four adults and seven doors, we had to sneak from one room to another while the children were performing and being entertained. The men put on a show for the kids and had them singing and dancing too. Our girls still remember it.

Another stormy Halloween, Dave and I loaded the girls into the wheelbarrow in rain gear and covered them with a tarp tent and wheeled them up to Estevan. We arrived to trick-or-treat our lighthouse neighbours. When they answered their door we said we had come as "West Coasters!" Supper and bunks for the night were provided, as it was a two-hour hike on the old plank road with five miles between us and the lighthouse.

GRANDPA HESQUIAT

Grandpa Hesquiat was the mail and freight man for the coastguard and for the keepers at Estevan Lighthouse, and when he retired, an old Davidson lifeboat, the workboat for the lighthouse, was left behind for him. He did off-loading for the *Princess Maquinna, Tahsis Prince, Princess Alberni* and other boats that carried passengers and freight up and down the coast. As he grew older and was no longer able to go on his own, we toured Grandpa around to beaches he was unable to visit alone. We took him to Homis, Cow Creek, the sand beach, Ray Basin, Cougar Annie's, and Hot Springs Cove. We used to chuckle when he would roll up his pants to his knees and use his cane to leap out of the boat and wade ashore, a Kodak moment! He would always find something to take home, sometimes a canvas of sorts for his next painting. Our girls loved to find treasures for him, too: bones, driftwood, corks, floats and glass balls.

Grandpa Hesquiat was injured in the tsunami of 1964 while aiding people in Hot Springs Cove. He developed a limp, which wasn't diagnosed until 1980 as a chipped bone in his hip. He was unable to walk far as a result, but was very proud of still being able to walk into the bay around the corner from our house. He would go there frequently just to sit on the beach.

My youngest daughter Korianne had a grade four school assignment naming a local place after an important person in her life. She never knew Grandpa, as he died before she was born, but she still felt his influence. Consequently, she named the first basin Hippolite Bay after Grandpa; that was his first name.

Grandpa Hesquiat used to visit his other children, sometimes in Kyuquot, other times in Ahousat or Port Alberni. He would be gone a week or a month and then he would show up or call for a ride home. Most times he would be deeply homesick by the time he arrived here and crave local food. I used to ask him what he

The hot springs at Hot Springs Cove. (PHOTO: JULIE COCHRANE)

Waterfall, Great Bear Rainforest. (PHOTO: JEN PUKONEN)

would like for supper when he got home. One day he wanted bird soup, so I asked what kind of bird.

"Any kind!" he answered. "Even one of those turkeys down on the seawall!"

He was referring to cormorants, which are highly unpalatable. Luckily I had some jarred mallard, so I made him duck soup, a dinner he really enjoyed.

We lived with Grandpa Hesquiat for nine years, until he died in December 1983. With his help, I learned many Hesquiat words, and we taught our girls to speak all we could learn. He enriched our lives in every way. His artwork still decorates our home and his picture graces our wall.

LIVING ON

Visitors and hikers often ask us what we do here all the time. They find it so isolated with its population of four. Well, we are obliged to tell them there is never enough time to do everything we'd like to do. There are always ample projects on the go. We've built outbuildings and cabins and cleared plenty of bush. Since Jody started kindergarten in 1981, a great deal of time is taken up teaching the children. As I home-school my kids, I've been through lots of curriculum changes and upgrades. The courses are becoming more efficient all the time. I even learned how to use the computer on which I am writing this account.

Our life here in Hesquiat has been a great learning experience for us all. When I look back, I can't imagine living anywhere else, or any other life. This is home, for me and my family, as it has been the traditional home for Dave's family for a very long time. There have been many innovations since we first came here, many changes to our remote lifestyle, and probably many changes to come. Life is what you make of it, and I'm glad for such a rich life as the one I've had here.

At some later time I hope to present the world with a longer reckoning of our life in Hesquiat.

A Sound Existence

- ROB LIBOIRON -

I consider myself fortunate that the definition of the word "home," for me anyway, has always been one of comfort and security, even if my homes have been a little unusual compared to most.

I first came to Clayoquot Sound some time ago, preferring the storm-swept beaches to city sidewalks, the roar of the surf to the roar of traffic. Being a nature- and especially a water-lover, the emerald sea of the Sound was paradise found, as yet unspoiled by what many people call progress. Those of us who call Clayoquot "home" do so proudly. We know the scenery here is admired from around the globe, and that this place, a Biosphere Reserve, has become a tourist Mecca. But I believe it is isolation that is needed to strengthen our connection to wild surroundings. If such a retreat can be called home, even better.

For three years, I lived on a wooden boat in one of the most weather-exposed locations around Tofino. The sea bottom there is littered with numerous wrecks. Old wooden boats demand a fair bit of maintenance to remain afloat, but they contain an aura of character that cannot be bought at any price in a more modern vessel. Various creaks and groans, although alarming at first, soon became familiar, coming from subtle self-adjustments in the rigging, or just from the whims of wind, waves and tide rushing past. My live-aboard home was a sixty-plus-year-old, slightly modified ex-troller with an assortment of other boats and floating docks all tied together and anchored en masse.

The water was a mirror of the sky much of the time. Sunrise and sunset were therefore twice as bright and colourful. Anyone living on land would have to have been standing at the water's edge to visualize or appreciate the same effect. As well, there was no shade from nearby objects, and a slight sea breeze usually kept most insects and other pests ashore.

Kayakers were everywhere, from solo paddlers to groups of a dozen or more and they passed by in all directions. As for wildlife, seals were constant companions as well as eagles, herons, cormorants and other fowl. During peak migration periods I could even witness the occasional passing of different species of whales from a comfy deck chair.

The lunar cycle and resulting tides played a large role in my life, especially during full moon periods, when tidal rise and fall became dramatic. Large rafts of drifting kelp, other debris and sometimes entire trees would succumb to the elements and be set adrift. There was always the hazard of such objects becoming entangled in my anchor lines. One calm, sunny morning I arose to find an entire sizable cedar tree afloat and pressing gently on my main bow line in classic T-bone fashion, held there in delicate, teeter-tottering balance by the tide rushing towards my boat. Luckily, my friendly neighbour, Pete, appeared and helped me pull it free. The tide carried it swiftly away. I hated to imagine the same thing happening in the middle of a dark, stormy night. Even on calm nights I often had to rise from slumber to go out and dislodge tide-swept stray driftwood from between the boats. Aside from the irritating thumping noises, the potential for lethal damage always necessitated immediate removal.

On a night in mid-December, 2002, Tofino experienced a winter storm in which hurricane-force winds were recorded at Lennard Island. Even my forty-foot vessel leapt about like a rubber duck in a bathtub. All night I was checking the tie-ups, having to traverse the pitching, rain-slicked decks without handrails. Conversely, bouts of raging weather became some of my favourite times to stoke up the wood stove and curl up with a good, stormy sea book for the added ambience. That winter season of 2002/2003 had more violent storms than the following two winters combined.

Some mornings, and many nights as well, Tofino and its foreshore would drift in and out of fog banks. Some would be long and feathery, while others were a solid grey-white mist that could drop the temperature several degrees and soak you unawares within several minutes. I used to love drifting off to sleep on the gentle Pacific swell with the Lennard Island foghorn hooting deeply in the distance, while the lighthouse beacon arc swept the sky, reflecting off cotton candy wisps of fog. That was the last winter of the foghorn, and even the lighthouse beam can no longer be seen in the sky from that vantage.

At the beginning of 2005, the opportunity arose to move eight miles from town to a small hourglass-shaped cove, the back half of which is shallow, containing an oyster reef that becomes mostly exposed at low tide. The front half of the cove deepens out, and that is where I maintain a gangly collection of rustic floating buildings and rafts. The whole cove is ringed by four small islands that protect my floating assemblage from the worst weather. From dawn till dusk, a glance in any direction from my multi-windowed floathouse is an ever-changing kaleidoscope of nature. Since I'm living on the water, the rise and fall of the tides produce stunningly diverse and beautiful "shorescapes" all around me. As well as the postcard vistas, there is the wildlife that abounds in the vicinity.

One frosty February morning, I gave slow-speed pursuit to what I thought was my fluffy grey tabby cat as she waddled her way across the slippery wooden surfaces, carrying something awkwardly in her mouth. Over dipping and bobbing icy gangplanks, from float to float, this comical chase ensued, the trotting feline staying ten to twenty feet ahead. Nearing the end of the last dock, I tasted victory, as there was nothing beyond but ice-cold water. The critter I was after, however, didn't even slow down. Upon reaching the water, it dove straight in and proceeded to swim to a nearby island. Only as I slid to a halt did I realize that I'd been pursuing a raccoon instead of my cat, and to this day I still have no idea what was in its mouth.

Raccoon families emerge from the forest and forage on the temporarily exposed seabed, concentrating on turning over small rocks. As well as eating tiny crabs, raccoons like to dig for shellfish in mudflat areas and have been known to suffer a high mortality rate during severe bouts of red tide. They also frequent my docks at night, feasting on mussels picked from exposed anchor lines. On quiet evenings, they can often be heard crunching away right outside my windows.

During low tide, bears feed on exposed grasses. Often, they'll lie down and stretch out, continuing to graze contentedly. They also love to roll rocks and lap up the numerous small crabs and blennies — scaleless, long-bodied fish — that collect underneath. Bears will just as eagerly scrape mussels from the rocks as an added staple. Whether grazing, rock-rolling or simply lolling about, these animals have lingered in plain view for hours, sometimes nudged back into the woods only by the rising tide. For two seasons, the main resident bruin was a female with two adorable cubs. Throughout the following weeks and months, I enjoyed watching their progress from berry-picking to swimming lessons. Bears are good swimmers, moving from island to island.

A large family of otters comes to visit once or twice a week, diving under the docks and climbing gleefully over everything in sight. Their chief diet around here consists of fish and the odd crab, but mostly mussels, which grow by the thousands under my floats. Even their scat — deposited here and there when they're visiting — consists mainly of ground-up mussel shells. When either amorous or aggressive, they can look and sound growlingly vicious.

Very occasionally they carry on loudly and vigorously under my floathouse floor. The first time this happened, my cats bolted for the loft and stared down, wide-eyed.

Swallows nest under the eaves and kingfishers chatter over territorial rights between dives into the water. Merganser ducks snorkel about energetically and produce some large broods in the spring, fluffy young ones constantly dipping and skimming about. When I notice one adult cruising by with over twenty young, all a-buzz around her, I know babysitting services are being shared.

On one memorable occasion, three wolves emerged from the trees, each one a different colour. They bayed and cantered playfully about the shoreline all too briefly before the forest reclaimed them.

Apart from wildlife observing, another enjoyable activity is driftwood collecting. There are miles of shoreline, but the best areas are dead-end inlets, where time and tide can leave vast deposits. I make great use of the more unusual pieces for decorative and functional purposes alike. Shore-cruising for driftwood is therapy for cabin fever, loneliness or general melancholia. The more vigorous activity of cutting firewood can cure anger and frustration. I also take time to cavort with my pets. Out here, mental health is equally as important as physical health.

As more and more people are discovering Clayoquot, finding room to be by oneself can become increasingly difficult. Those who choose remoteness as a lifestyle may experience raw nature as few others can; there are, however, many ongoing hardships in this kind of existence to consider. Life's usual comforts are not there at the flick of a switch, and I must bear in mind that even though I may be sensible about safety and health issues, help can still be a long way off.

A couple of years ago, I found out the hard way that I'm epileptic. Luckily, my sudden seizure occurred in the company of others. I was subsequently prescribed a remedy; as an added precaution, I voluntarily have monthly blood tests to ensure the level of my medication is appropriate. Despite my proactive measures, however, there are moments when the dreaded pre-seizure symptoms en-sue. More than once I have wedged tools under my doors to block them from myself, in case I might wander during the blurry state following a seizure. Once, I felt compelled to write a note in case of my demise. It was addressed to no one in particular. After years of living remotely, I have come to accept that I am just as mortal as the animals.

That doesn't keep me from worrying, though. With an eye on the barometer and an ear to the V.H.F. marine weather reports, I have many anxious moments, especially in storm season. During the late fall of 2006, coastal B.C. experienced tempests of a magnitude seldom seen. While large tracts of trees toppled in Vancouver's Stanley Park, I had no rest at home as again and again I had to step outside and face the elements threatening my homestead. My main anchor lines held, but secondary lines and gangways snapped as swells rolled into the bay.

Indoors, things were thrown about. A lamp fell and shattered, breaking my toe and leaving glass embedded in my foot. At one point, a kerosene fire made things even more demanding. As the storm continued to rage, I began to wonder what more was in store. Eventually, however, the weather calmed and I was stuck with my injuries, a hefty tidying-up job and numerous repair projects. Since that autumn, whenever an approaching deep low sends my barometer needle plunging, I tap the glass and wonder how much this one will hurt.

I still keep a boat moored in Tofino's harbour, in case inclement weather or other circumstances prevent me from making the trip home. Now, more than ever, I have to live my life by nature and her elements, mainly wind and tide. If these two are pitted against each other at any time, and especially during full moon, motorboat travel between town and home can be rough to impossible, with miles of high, standing waves and a fast-flowing current. The half-hour trip is more often a pleasant experience, though, and aside from the obligatory seals and eagles, porpoises and even whales have accompanied me.

A short time ago I had a very close encounter with an adult bald eagle on the small aft deck of my boat near town. Emerging from the cabin doorway, I suddenly halted as I became aware of this huge bird of prey standing about four feet from me. It leaned instantly away, but showed no sign of flying off in a panic. For perhaps a full ten seconds we studied one another closely. We were so close that I could see its chest rise and fall with each breath. I marvelled at the keen intelligent gaze I received in return. The early morning breeze ruffled a few of the huge bird's feathers, and then the eagle crouched slightly and heaved itself into the air. No matter where I hang my proverbial hat for the night in Clayoquot Sound, its wild inhabitants often make their presence known.

For now, here I stay. This life is not for everyone, but I feel privileged to call the wilderness home and thereby embrace the wildness within.

Lingering

Stone Heart

– ADRIENNE MASON –

The single whistle of the saw-whet owl pierced through the dark, startling me just as I was drifting off. This note was the first of hundreds that would resonate out of the black throughout the cool May night. At first a novelty, the call soon became a monotonous serenade, the metronome that measured our work. Bundled as best we could in layers of fleece and rubber, my companions and I waited quietly in the forest, sometimes near motionless for over an hour, until a frantic fluttering in the forest signalled it was time to move and tend the task at hand — weighing and banding the chicks of ancient murrelets.

I was glad for the saw-whet's first call. The cool earth had sapped my body's heat and I'd stopped feeling my toes long ago, my rubber boots being completely inadequate for the spring night. It was a bitterly cold evening, unusually clear for the west coast, especially in the Queen Charlotte Islands, the misty isles. The usual blanket of low cloud, that kept the climate drippy but mild, had peeled back earlier that day and now, at one in the morning, stars peppered a clear sky. There were stars in the sea as well. When we'd putted ashore earlier that night, navigating by the light of the moon, bioluminescent plankton agitated by the action of the spinning propeller had trailed our boat like a stream of liquid glitter.

I treasured this time in the islands. Before I had children, I was able to visit once or twice a year and was lucky enough to be paid to do so, as a guide on sailing charters. Now, with two young daughters, trips were infrequent and all the more precious to me. Being outdoors is central to my life; it fuels, restores and, in so many ways, defines me as well as my relationship with my husband. We had met kayaking, on a warm, summer Rocky Mountain evening. Our courtship involved more skis, bikes, tents and scuba tanks than fine dining and flowers. Wearing a forty-pound backpack and a filthy, sweat-drenched t-shirt, Bob proposed to me with a cherry-sized chunk of quartz from Jasper's Tonquin Valley. We were at our best as a couple when we were outside — our ceiling, the stars; our carpet, a mountain meadow. It seemed there were no walls.

After our marriage, our lives continued much as they had before. We both had jobs that allowed us to be outdoors a lot. Time off meant more exploring. We were short on money, but flush with experiences. We lived the creed of Edward Abbey who, from a yellowing scrap of paper taped to our fridge, encouraged us to "get out there and mess around with your friends, ramble out yonder and explore the forests, encounter the grizz, climb the mountains. Run the rivers, breathe deep of that yet sweet and lucid air." We were happy to oblige him.

Then one day, a few years into our marriage, Bob got all responsible on me. Considering he's ten years older, I guess it shouldn't have come as a surprise, but still, I just wasn't ready. Buy a house? Think of all the trips we could take with that money? We had too much to do before settling down. His response? I swear it's true — "You can't eat memories." No, you can't, but the idea of filling my mind and photo albums with images of foreign lands and bagged peaks sure tempted the taste buds more than a mortgage we would surely choke on. And the thought of kids? Didn't we actually need some furniture first?

Despite my relentless teasing of my older husband ("It's your birthday dear, why don't you take the day off? Call in old.") this was the first time that our age difference had really came to the fore in our relationship. Of course, in hindsight, we *both* seemed laughably young, but as a woman just twenty-eight, I struggled with the thought of conforming to the stereotype, of having a mortgage and a child so soon. Could a mini-van be far off? (It wasn't, as it turned out.) I loved my job, I loved our freedom; surely there was a better — read further off — time for such responsible behaviour.

But I am weak when it comes to the sensible pleas of my earnest older man, and before long we were the owners of a modest stuck-in-the-seventies bungalow, complete with harvest gold appliances, fake-wood panelling, and wall-to-wall brown shag (oh, hallelujah, even in the kitchen). Bob's rational arguments swayed me on the merits of making the biggest, and frankly most uninformed, purchase of our lives before I was thirty. (And of course now, fifteen years on, writing from our "retirement fund" — our heavily renovated, ridiculously over-assessed home in Tofino — I am forever grateful.)

Not surprisingly, kids were next on the list of really-important-things-that-were-going-to-cost-a-lot-and-cramp-my-style. We always assumed we'd have kids, but, again, it was a matter of timing. The chimes on my biological clock hadn't even started to emit feeble pings while Bob's were clanging like a three-year-old let loose on a drum kit. While I was trying to calculate the best time to get pregnant so that I could work as long as possible and still fit in a few sailing trips, Bob was doing the math on how old he'd be when his child — assuming one was born that year — would graduate. We're both lousy at math though, and despite my best attempts to orchestrate the perfect timing for a debut, our first child was soon on her way.

Suddenly, it seemed, we had our nest and our chick.

Out of the inky darkness a murrelet whizzed past my head and flew into the forest. An adult was homing in on its nest and its chick. Ancient murrelets are true seabirds, only coming ashore to breed and nest. Months before, each pair had dug a two-metre long burrow into the island's soil, and the female had laid a pair of speckled white-and-brown eggs. For a month, the parents took turns warming the eggs by snuggling up to them with their brood patch, a featherless spot of skin on their belly. Finally, after this period of intensive care, the chicks started to chip through their shells. About six days later, they would break free of their calcium prison. For many of the island's chicks, tonight was that night.

Biologists describe newly hatched birds as being either altricial or precoccial. Altricial birds are essentially helpless — eyes closed, devoid of insulative down — and need intensive parental care. In contrast, precoccial birds are capable of moving around and keeping warm immediately after birth. Ancient murrelet chicks are famously precoccial in the avian world, being able to feed themselves from the moment they hatch. Within two days of emerging from the egg, these chicks head to sea, little balls of grey fluff scrambling and tumbling along the forest floor on adult-sized feet.

Our nocturnal visit was timed to intercept the seaward cross-country dash of the chicks. Earlier in the day, plastic fencing, about a foot high, had been set up outside the burrows with active chicks. The fencing funnelled the birds down to several stations where volunteers such as myself waited in the dark, listening intently for scuttling in the tunnels. When we heard the birds we shone our light on them briefly, gently placed them in a cloth bag, and carried them to the banding station. There, the chicks were weighed and given a silver anklet with a unique alpha-numeric code, denoting the place and time of their capture — the murrelet's equivalent of a certificate of birth. We'd then return the chicks, so they could resume their scramble to meet their parents who were calling for them just off shore.

All of this silent waiting in the dark got me to thinking about my

Anchored in a quiet bay, Gwaii Haanas. (PHOTO: JEN PUKONEN)

Nature taking over. (PHOTO: JEN PUKONEN)

own two chicks at home. Yes, seven years on, we'd miraculously managed to keep up with the payments on the house to which we'd added two beautiful daughters and a motley assortment of second-hand furniture. Despite my thought that somehow I'd do my life differently and I would never conform to the "norm" (whatever I imagined that was) our lives and our roles, as parents and partners, were pretty traditional. While I tried to freelance on occasion, I was the main caregiver to our girls while my husband worked full-time. I knew I had it good — a great relationship, two wonderful daughters, a decent home, the occasional interesting writing job, good friends and a husband who was active not only in raising his children, but who was also a much better housekeeper than I. Still, it was so hard to adjust to not being able to do what I wanted when I wanted to do it. I cringe when I remember the supreme hissy-fit I pulled when Bob called from work saying that he'd be home just after his run on the beach. Damn it, I couldn't go for a run whenever I chose, and I resented that he could.

Slowly, thanks largely to Bob's unending patience, we navigated our way through the peaks and valleys of a new reality: life with children. Of course, we also wanted to tackle a few real peaks and valleys. Having kids didn't squelch our wanderlust, but we learned to shift our perspective and scale. We still hiked, camped and explored, but it was no longer always possible (or even desirable) to get to the end of the trail or the top of the mountain. Instead, with one child on my back and another's hand firmly grasped in mine, tide pools became oceans to explore; ditches, rivers to raft; boulders, mountains to climb; and mud, just something a whole lot of fun to wallow in.

As the children grew older, and my younger daughter was weaned, one of the few paying jobs I could occasionally fit in — with a lot of help from Bob, family and friends — was to work as a naturalist on sailing charters. It was the perfect "job" — a short getaway to somewhere stunningly beautiful and so, so quiet, with a boatload of interesting people, and a modest paycheque waiting at the end. So this is how I found myself shivering in the forest, listening for murrelet chicks blundering in the runways, so privileged to be here, yet desperately missing my kids and mired in guilt for having left them.

On this trip, my angst was particularly prevalent because of a message — cryptic or otherwise, I'm still not sure — that my seven-year-old had sent me just before I left for this trip. I was frantically trying to meet a writing deadline before rushing out the door to catch my bus to Vancouver, and then a plane to Sandspit. Ava was hanging around my home office, wanting to be with me, realizing much more than her younger sister, Patrice, that I'd be away for a few weeks. To keep her busy I asked Ava to draw a picture of our family on the front page of the new journal I was taking on the trip. Ava quietly took to the task, and when she presented it to me, I briefly glanced at the picture, gave her a hug and said, "It's beautiful honey." Perhaps I even added an offhanded, "I just love the colours." Whatever I really said, I hadn't looked at the picture closely at all.

But earlier this morning, when I had finally cracked open the journal to write a few notes, I had a better look. The four of us were in the picture, all with the right number of body parts, my eyes were bluer and Bob's beard was redder, but we were all there holding hands. But while Bob and I were smiling in the picture, both girls had absolutely straight lines for mouths. No smiles, not quite frowns, but there were definitely no upward curves in those red Crayola lips. The picture was an art therapist's dream.

All mothers navigate through new emotional terrain once they have children. And it's seriously unmapped. Sure, there are mountains of help-thyself-and-baby books, and armies of women willing to give you — free of charge even! — their advice, but nothing *really* prepares you for that first child. When Ava was born I didn't feel an overwhelming desire to see her at first. I held her for a bit, but I remember feeling relieved when she was taken away for . . . well, I actually don't know what they did with her, but I was okay with that. At that moment I think I was stunned into a sort of mild shock. Now, in fairness to me, this particular birth really had come as a shock. Ava blasted into the world in just less than four hours (the fact that the receptionist phoned the nurse's station to tell them "We've got a live one here!" might give you some idea of what was going on in the waiting room) and she was two weeks early. I had deluded myself into thinking I still had weeks to get my head around the impending birth, but no longer. Baby had arrived.

I haven't been to anyone else's birth, but I've seen enough pictures, movies, and commercials to be led to believe that most new mothers are sobbing over their newborns and hugging everyone in the room moments after giving birth. I was happy of course and she *was* the most beautiful baby, and, truly, I do think birth is a miracle, but . . . I hate to say it . . . I just needed some time on my own, and I felt somehow that wasn't an appropriate feeling given the import of the moment. My baby wasn't even an hour old and already I was feeling guilty.

Looking back, I think this moment gets at the crux of my biggest challenge with motherhood: sometimes, moms just want to do their own thing, but when we do (or at least when I do) we miraculously find a way to feel bad about it. I'm not sure why this is. Certainly no one else was making me feel guilty; it was self-imposed. In parenting, there is no end of things to feel guilty (or at least completely inadequate) over. Everyone has an opinion, there is a book, or five, to address every angst a parent might feel. Since I don't watch a lot of TV, I can only imagine the advice that must spew forth from there. Parenting firmly lies in the too-much-information-is-*not*-a-good-thing camp.

And so we made our way, husband, baby and I. Hours after Ava's birth, I *was* cuddling that beautiful baby, staring at her, stunned, amazed and scared shitless that she was actually mine. Life-with-baby changed so much in my relationship with my husband and, well, in my relationship with myself. We navigated the terrain as best we could — guided purely by gut. There is such joy being a mother, but it's scary, too. You only get one shot as they say, and the pressure's on big time not to screw it up. It's also really hard work trying to keep an eye on what makes you happy *aside* from being a mother. You're trying so damned hard to do a good job, it's easy to lose track of what makes you, you. Like most families, we muddled through, and we found the rhythm that works best for our particular quartet. Most of the time, of course, it is the kids' needs that come first, but we've also learned that there are times when it's okay to put the needs — okay, maybe they're just desires — of a couple, or even an individual, first too. Through it all, you just hope like hell you're getting it right.

Around me, others were also silent with their thoughts, disturbed only by an occasional chick in the runway and the whistle of the saw-whet, setting the constant, rhythmic backdrop to our nocturnal watch. Periodically there would be a flurry of sound in the forest as an adult murrelet flew in low, homing in on its burrows. The adult would perch at the lip of the burrow and call to its chick, making what is called an imprinting call. The chick would reply and a period of intense vocalization followed as the parent flew back out to sea to wait for the chick, which would begin its solo overland foray into the forest and towards the sea. The desire to follow their parent is intense and innate. While the descent to the sea was fairly gradual in this study site, it is not unheard of for murrelet chicks to leap off 200-foot cliffs into the surf below in the instinctive drive to reach their parents.

I thought of the murrelet chicks and the calls they made back and forth with their parents. If the chicks vary their path and wander off track, do the parents call more intently, signalling with vigour, "I'm here, follow my voice. I'm here, follow me." Do the adults know the smell of their chicks, as I know the smell of mine? Can they recognize their cries from across the playground? Do they hold their breath when their chick makes the plunge from land to sea?

What a leap of faith that is, how much like entering the ring of parenting — this leap of faith we take with our partners when we create a new life. The leap of faith we take as we teach our children and guide them through a world we hardly understand ourselves. How do we ever know if we've got it right? Have we taught them the things that truly matter in life? And more to my immediate angst, was Ava subconsciously telling me something with her drawing, or was I adding the message, fuelled by guilt and self-doubt?

I thought of the past Mother's Day, which had occurred just

before I left for this trip. Ava had given me a small paper box that she'd cut and folded and pasted together with her first grade classmates. The box was covered with flowers, a sun, bees and butterflies. Inside were messages she'd printed: "I like it when you ride bikes with me," "I like it when you cuddle me," and "I like it when you tuck me in." Such simple things; such simple needs. Along with this gift, she'd given me a heavier package, wrapped in yellow tissue. Inside was a rock shaped like a heart. Of course I treasured that gift, but what I loved even more was that my children would know that a stone, or a feather, or piece of beach glass softened by the sea, was *exactly* the kind of gift that I would want.

That night on Limestone Island, we captured and banded seventeen ancient murrelet chicks. At about 3:30 A.M., we packed up our gear under a shimmering sky streaked with northern lights — the first I'd ever seen on the coast — and headed into the forest for a cross-island hike back to our boat. Cold and tired, we welcomed the chance to move our aching bodies, yet we all remained silent, deep in our thoughts, as we navigated our way through the dark, stumbling over tree roots and squelching through pools of mud. I led the procession and emerged onto the cobblestone beach where we had left our boat, just as the moon became perfectly aligned with the tiny beach. The orb of the full moon loomed above and bathed the beach in a brilliant white light. In the middle of the beach, a rock stood out from all the others, caught and glowing in the moon's wash. It was a palm-sized rock, perfectly shaped into a heart. I picked the stone up, warmed it in my hands and held it against my cheek, wave-washed, smooth and surprisingly soft.

A Series of Poems on the Paintings of Emily Carr

– KATE BRAID –

EMILY CARR'S *TOTEM MOTHER*

She wears no ornament of any kind.
Only a lap as ample as any wooden bench,
shelter under a wide brown chin.
I hide in the cool fibre of her arms,
her child, yet
she dreams away from me, a future
that hangs suspended, shimmering
on the edges of a coming night.

Beyond this shelter, this cave
of light and mother's arms
is another place.
Mother longs to go there now.

When she lays me down I shall feel
the soft, wet crinkle of grass piercing
the tops of my thighs.

I'm not sure I want to leave, no,
but there is that look. Already
her grip loosens, inexorably down
toward the bright green
of the forest floor
I begin to slide
caught
in the terrible
single moment

Mother, mother!
Catch me!
I fall!

EMILY CARR'S *WESTERN FOREST, 1929–30*

She watches   waits.

This weight on her chest, abundance
          smothers her   hovers
in the disguise of forest.

Here are hidden moving feet.

And there! That bent figure
          in the undergrowth — she caught it,
doubled over, laughing at her,
us.   Ominous

but she stays
stubborn. *This is forest* she testifies.
And look, she brings gifts:
two small tubes of green paint, one of brown.

Trees, bushes, undergrowth, all
buzz at her cheek, the nerve of her and yet

she will not be driven out.

Undergrowth snaps its outrage.
Leaves brush leaves in a sibilant hiss.
Branches crackle, tree trunks
(their muscle)
wave and sway from darkness.

Here lies shadow no wind only
tumultuous green, furious brown
and small round Emily

below just painting
waiting
for the invitation to come in.

EMILY CARR'S *LOGGER'S CULLS*

She paints
another landscape wiped bare
by careless human hands —
here it's loggers' culls,
there, a city.

Something has died.

Only the clouds are alive, one stand of forest
and there, to the left,
a crystal ball of ghosts.

The sky knows. It ties itself
in a tight blue knot.
Whirlwind with no way out,
it circles, keening.
At its centre shines
a deep blue eye.

EMILY CARR'S *STUMPS AND SKY*

Today as I walk through another of those desperate gardens
left by loggers passing
with their god-awful axes, saws, knives

I feel lost.
I can't bear it any more, green and beauty
all slipping by.

Death deals the most colour,
a solid black in this deck of falling cards.
I should never have strayed
from the kitchen. Domesticity at least
paints firm lines, sets out fences.

Here in this fading larger world
I begin to lift, ghost-like
from the pale, pale ground.

EMILY CARR'S *THE RAVEN* (MAUD ISLAND, HAINA, QUEEN CHARLOTTE ISLANDS)

This raven, throat distended, sips wood.
Urgent, it is willing, will risk anything
to consume   more
of the thin dark line
it already is.

Two pine trees below
tip their heads, timid, in the bird's direction.
Take me, they say, afraid
to be caught wanting and yet
so thin, so dry. They wait for the black beauty
that forces itself, that would feed on their willing bodies.

It is their only hope for flight.

The bird has not yet reached that height where it yearns to be.
(More wood!)
Someone has carved its wings too thin
too fragile   to lift.

In its haste to rise
it swallows everything, strains
upward, eyes squeezed shut

it doesn't notice
    the sky is already blue,
        the winds fair
pine trees
     waiting.

EMILY CARR'S *GREY* c. 1931–1932

Since painting "Grey" my seeing has perhaps become more fluid.
I was more static then, and was thinking more of effect than spirit.
— Emily Carr, *Hundreds and Thousands*

So, when the shoe fits
The foot is forgotten. . . .
Easy is right.
— When the Shoe Fits, *Chuang Tzu*

Is there anything so solid?
It surprises you, how it turns up
when you're not looking.
A glimpse of what holds the whole thing up.
Call it core. Call it bearing beam.

Some are so grateful, they call it salvation
then worry they're going to be crushed.
Some try to sell it.
Others just keep on walking.

Do you sense straight lines here?
Anything hard?
Easy is right. Press closer.
Try circle, try rising.
Try arms. Heart. Eyes wide.
Try open.

EDITORS' NOTE: *Kate Braid's* To this Cedar Fountain *(Polestar, 1995, Vancouver) offers another series of poems on the paintings of Emily Carr.*

Hornby Island

The Nature of Home

– KEITH HARRISON –

On this island that has the shape of a wonky boomerang, the new day always begins with wood. Dim forms of trees reach out from the charcoal sky, stretch up, and stand. I get up and wash my face in the cool water of the sink set into the countertop of glued-up, finger-width strips of alder — generally considered scrap wood but here the diverse grains and alternating hues disclose an unexpected loveliness.

With a cardboard box of cereal in hand, I step over an inlaid line of black-brown eucalyptus. It offers a warning contrast to the once-whitish beech flooring which the sun has now turned a warm yellow — and marks the descent into the low-ceilinged dining room. Seated at the round table of bird's-eye maple, I look out

---

Part of this material in a slightly different form appears in Elliot and Me *(Oolichan, 2006).*

beyond the untreated curved cedar deck to a very large, very graceful arbutus that a University of Victoria botanist said was here before Columbus. Like Michelangelo's *David*, it is so aesthetically pleasing that its larger-than-life scale continually surprises the eye. On the edge of a cliff, rooted in conglomerate rock, it leans south out over Tribune Bay, a connecting body of water that is (like myself) exposed to distant influences both south and east (my father, a student in Melbourne, was born in Tokyo). Hornby Island, itself, also began in the southern hemisphere, one of the ancient volcanic islands of Wrangellia, which with geologic slowness rode north across the equator on the Pacific Ocean plate.

Half a dozen mature second-growth conifers are visible in the foreground as I munch away. Their lower trunks rise branchless, echoing the round barkless posts that hold up the squared-off beams of this wooden and glass structure that owes as much to the homes of the Haida as to the spareness of Japanese architecture. The Douglas firs, reminders of the external living origins of my house, also point to my dead father's work with books and newsprint as a teacher and a journalist, and to his later job for a forestry company that, for marketing purposes, rebranded hemlock as Alaska white pine. But it is to the magnificent, uncommercial arbutus tree my gaze again and again returns. Of no interest to loggers, this hard tendony wood keeps on twisting and flexing even after it is felled. Arbutus has only the drama and uselessness of beauty.

To my left are a couple of couples: examples of arboreal intimacy. A dark, rough-barked vertical fir is squeezed in an ecstatic arbutus embrace — whose two smooth red limbs are flung exuberantly upwards. It is a spirited reminder that I do not live alone but conjoined, and in the decades-long sharing of habitat and touching and coupling and growing together and supporting of each other, we are not unique. Nearby there is a second, smaller pair. Edging over the 200-foot escarpment, a squat arbutus wraps its arms around a lean fir whose roots — due to erosion of the stony cliff-face — now dangle mostly in the moist air above the Pacific Ocean.

I cross a cedar walkway to my writing studio because, as a novelist, I'm a believer in the Scottish proverb that an hour in the morning is worth two in the afternoon. Inside are many books, a computer, and a large box of paper; outside the window is a grove of slender arbutus trees — kin to the gums of Australia? The ocean shifts below. Gradually, half-alive prose appears on the grey rectangular screen. By late afternoon, if I'm fortunate, paper has become a few inked pages that might become a book that might require the taking down of a modest, non-old growth tree somewhere for a print run.

When I walk out the wide front door, often with my wife, I look at the garden that is mainly grasses but has two Japanese maples — one green, the other red — and a ginkgo tree in a large pot. None are native, yet they relate to my sense of a familial past. However, they do not flourish because deer munch the leaves for their early breakfast. Not liking fencing or netting, we have tried to discourage Bambi with bars of manly Irish Spring soap and tea bags of "panther piss." Perhaps this rotten egg spray will work. The path off the driveway leads to the most northerly stand of Garry oaks in the world. All of the same advanced age, these spindly trees with their curled, dusty-green leaves pose a riddle about beginnings — and perhaps foretell an ending — since deer browse to death any young ones. Were these oaks born into a pre-deer island world? Was it some weird phenomenon of predator satiation in which the sky rained acorns too numerous for the appetites of quadrupeds with deciduous branching horns?

As we continue along the cliff path, my lungs swell with woodsy salt air. Cities are indispensable for me, but for too long — living in Vancouver and Montreal — I've breathed diesel fumes, dingy clouds of half-burnt gasoline, and the tarry haze of asphalt. In the spring, I pause to stroke the bendy softness of the new, pale-green tips of fir. The older, darker needles feel stiff against the pads of my fingers, but the trail to the Helliwell Bluffs ahead opens out to verdant grasses, and the ruffled sea beyond. On the left is a large arbutus tree that holds both yellow and new leaves in its wide, shapely canopy. Even when shedding, the sinewy limbs are never bare. The warm meadow is dotted purple with wildflowers, has glints of white and dabs of buttery gold. Where the forest meets the meadow are orangey-red streaks of Indian paintbrush. The broad outward hook of the peninsula shimmers with soft, undulating grasses, and occasionally a juvenile eagle with mottled feathers floats past.

Underfoot, I kick loose a round brown rock from the geologic mud pie. A wind-bent shrub has on its exposed southeasterly side only a few stunted boughs, but the other half is fully green. Looking down the Strait of Georgia to the far horizon on a very clear day, we can glimpse a bright white triangle that vibrates against the sky near the American border: Mount Baker. A strong whiff of guano comes off the cliff, and the barking of seals (or sea lions) grows louder. Torpedo-shaped cormorants wing in from the ocean towards the upthrust wedge of rock we're walking along. Like a postcard, the long meadowy vista framed by sea and forest invites us deeper into the stretch of wild grasses washed by the light. Moving along the bluff's down-and-up curve, we pass bare black branches and evidence of fire. There's an abundance of young pines that must thrive on ash. I inhale the sweet air, taste the echoing blues of sea and sky.

Off Flora Islet, with its steeply sloping gravel beach, there may be fishing boats. At low tide, Flora nearly joins Hornby, yet one narrow, fast-flowing passage keeps them forever apart. Leaving behind the ruckus of seals or sea lions from the reef beyond the islet, we follow the bend of the trail around Helliwell Point: now halfway back. A single fruit tree grows out of what looks like sea rocks, its leaves like thin hands. In the shallow water, a blue heron stands on one leg in Zen-like stillness. The professorial head gazes down from the long neck. At a school of fishes? All at once, in a laboured beating of creaking wings, the heron takes flight, its head tucked back on an S-shaped neck. As the strange creature flaps towards the next bay, just clearing the water, bent bamboo legs trailing, the subtle blue of the feathers fades almost to grey.

The path loops away from the beach into the forest. There is very little underbrush here, with wide avenues of light between the giant trees. These old growth firs have white-green lichen on their deeply ridged bark, and one holds up a huge eagle's nest. Further along, off the main path, on what could be a deer track, is a freak of nature. Two large cedars have grown together like Siamese twins. A thick lower branch reaches through empty air and links the two living trees. I can't tell from which round smooth trunk the branch originated, but one of these trees must have taken a stranger's limb right inside itself once. Like lovers.

From here the trail returns to the bluffs and the open meadows that look out on the mountain ridge of Vancouver Island. The walk back is westward, a retracing of steps to the large lantern of wood and glass resting on its side that it is our luck to call home. After dinner, as the earth rolls towards darkness, the cliff-dwelling arbutus tree that was here 600 years ago catches reds and mauves in its arms. And holds on.

Forty Kilometres from Home

– GREG BLANCHETTE –

Her name was Sara and she barely existed in this town. I'm not kidding — I reckon she must have been nearly invisible. Her number appeared in the Ucluelet phone book but her telephone went for days and days without ringing. Even when it did, odds are it was a telemarketer. They often had a nice long chat.

When she moved through this town she always walked, and she was always alone. Sometimes, rarely, she met other people out walking, and sometimes they even recognized each other and said hello. They'd comment on the tourists or the weather, or something else they were unlikely to disagree on. It seemed for a moment she flickered into existence. But then these acquaintances went on their way, and she realized it was an illusion. Like the

EDITORS' NOTE: *Greg Blanchette's story is a metaphorical account of lived experience.*

starlings flitting in the trees, like the Pacific lapping the edges of town, like the pathways disappearing into the forest, she was hardly there — something to be avoided or walked around, perhaps, but no more noticeable than that.

In her perambulations she learned that nobody else walks in this town — nobody old enough to buy a car, anyway. After dark she had the roads to herself; she could walk right down the middle if she wanted, and usually did. To keep herself company she took to shaking a tiny pair of maracas, little plastic bulbs with beads inside. At first, she liked the rhythmic swish they made in time with her footsteps, and the concentration it took to coordinate her hands. Then she started to like the air of eccentricity they gave her. Then she heard that rattles are an ancient instrument of power, used past and present by shamans the world over. They signify change, transformation. She began shaking with a purpose.

Soon she began noticing strange things. People in houses would look out secretively as she rattled past, then throw the curtains open as though there was no one there. When she crossed paths with people weaving their way home late from the bar, if she stood silently at the roadside, they'd walk right by. Only if she shook the rattles would they stop, jerking their heads back. "Oh!" they'd say, giggling drunkenly. "We didn't see you."

She began to imagine she was sowing transformation, walking the streets invisibly after dark, shaking the rattles. She'd imagine she was initiating conversations with the spirits that inhabited the ditches and culverts, waking them gently so that one night soon they would talk, really talk, and not feel obliged to steer cautiously away from what they actually wanted to say.

She would tell them she could hardly believe how lonely a human being can get. She'd say how ashamed she was about this, because all around her people were apparently having normal social lives, so why not her? Weren't people supposed to be social animals? She would say she had been led to believe that small towns were friendly places — that in retrospect, and though she didn't understand it at the time, much of the reason she left a great job and moved away from the city was for a word, a concept: *commu-*

nity. She yearned to live in a place where it would be easy to meet people. She dreamed, unconsciously, of a circle of people who knew her, of easy acquaintances, even friends. A place where she did things with people, rendezvoused on the beach for fires and potlucks, had people over, dropped in on them spontaneously, where they would call each other up to talk, like so many women seem to do so effortlessly. Alas, she was not like so many women.

Of course she postulated reasons:

1. *I need to get out in the community more.*
2. *I'm fucked up.*
3. *I don't click with the working-class demographic.*

Naturally, she also proposed solutions:

1. *Join classes — beading, cooking — or even play softball.*
2. *See a shrink and get some pills. Attitude adjustment.*
3. *Move somewhere else.*

Unfortunately the solutions required personal change, which put them out of the running, and that was the essential dilemma. Should she have to change to accommodate the place? Or should the place take her for who and what she was? The other question, an unavoidable one, centred on her: *Am I putting out some vibe that repels people? Why don't they call? Why don't I call them? Why don't we seem to have anything in common?*

As her isolation grew, and her ability to do anything about it shrank, the practical answer to those questions became "I don't know." Except for the ultimate solution, *moving away* — so often the answer (or at least the approach) to life's woes. After a few years in Ucluelet, her eyes unavoidably began turning to a place she thought might make her visible. A legendary place, a paradise to some, a grail to others: Tofino, just forty kilometres down the road. In Tofino she would finally feel at home.

Tofino casts a long shadow and, like me and my fellow Uclutians, Sara lived deep under it. Being down the road from Tofino is like living in the apartment next door to a high-class hooker, all ringing doorbells, fabulous dresses and strange men in good suits calling at odd hours. And sex-sex-sex. Or so you imagine as you sit home in your bathrobe watching TV, eating granola for supper. You know that the neighbour's life can't possibly be as glamorous as it seems, but still, you wish it didn't make yours seem quite so boring.

A while ago Sara came across an aphorism and it stuck like spruce gum to her brain. It was, she believed, from a Wallace Stevens poem, something to the effect that people live not in places, but in the description of places. It was true. Some mornings she stepped out her front door into what seemed like a broken-down backwater, surrounded by dying seas, with a backdrop of brutally stripped mountains, the tallest capped by a malignant radar dome pissing microwaves on the town day and night, and she thought, *it's no wonder nobody walks in this hellhole.*

But sometimes, especially on still mornings with the first fingers of fog probing the harbour, the water like champagne, and that sumptuous early light fanning up behind Mount Ozzard, it was a privilege just to be alive in the place. Home sweet home indeed. A dozen other small towns lurked outside her door, all with the same buildings, the same scenery, but all different, depending on the state of the weather, the time of year, how her cabin fever was doing, how she'd slept and what the day's prospects were.

If a town is more description than geographical plot, then Ucluelet — Ukee — has some serious P.R. work to do. Really, the joint hadn't done *any* describing of itself, not for posterity's sake, anyhow. Tofino, on the other hand, gummed together its collective myth of surf and coolness and extreme environmentalism and blared it from every media outlet it could find, in every medium it could tap. It became the world's desire, all thanks to the power of description. In a minor way Clayoquot Sound colonized the zeitgeist, and now every small town in the world was somehow judged against Tofino, and usually found wanting — no place more so than Ukee.

Sara had enjoyed friends, of a sort, when she first moved to Ukee. Before she went entirely invisible. Her last boyfriend had then discovered Sara's secret, and moved on, furious and humiliated, the big lug — but not before he told a few key people who made sure the whole town knew. Her friends were remarkably good about it — creeped out a bit, and curious — but they eventually moved away nonetheless, in pursuit of jobs and relationships and schooling. Several moved to Tofino.

But Sara didn't move. Call it bullheadedness, sloth or misguided loyalty to a town in which she barely existed. Call it whatever you want. She hung onto her housekeeping job, making other people's lives neat, working mostly alone. In the evenings she stayed home, alone. She spent weekends largely by herself. She hung on, walking, rattling, pining. Always with this vague, swelling feeling that life must be waiting to begin again, maybe just a scant forty kilometres down the road.

According to our joy-filled culture, there's nothing more useless than a fifty-year-old overweight woman without kids or a career. What's she good for? What can she contribute to the gross domestic product? What leverage can she possibly offer to the movers and shakers? Nope, she's a factotum — a sop for the dirty jobs other people don't want. And if she's something of a freak on top of it . . . well!

Do you know what it's like to be so lonely you feel hollow, as though you'd ring like a bronze statue if somebody were to hit you with a stick? Yes, sure you do; there's nothing unique about Sara's loneliness in all of human experience. Isolation, after all, is a theme of postmodern life. But in case you've forgotten, let me remind you: it doesn't feel good, which is why so many of us spend our lives searching, or maybe just yearning, for that "soulmate" to fill the unfillable void.

She started a journal because there was nobody else around to moan to. *Repeat after me: I'm unworthy. I'm a mess. I can't, I shan't, I won't. It's all over. There is no joy. But I still want, I still hope, I still yearn. The rope of my desire is anchored to the unyielding rock of despair. Nobody loves me. The fates are against me. My personality is an irredeemable mess. I repulse everybody. I can't hold a relationship together. I was born to solitude. Life is hard and getting harder. My negativity reinforces itself. The fun has gone out of it. This town is purgatory. Everything is forever.*

She made up personal ads that were never published: *Desperate misanthropist with flagging sense of humour wishes to hustle solitary stallion with taste for lit and handcuffs.* Some she did post on the Internet, an anonymous requisition for human contact: *Social misfit nonetheless yearns for like companionship. Age, sex unimportant but must possess ineffable attitude.* Nobody answered. She herself wouldn't answer such an ad. Resigned, she gazed out the rain-studded window and crumpled in on herself.

How many times had Sara stared out this window, or some other window in town, tears raining down her cheeks from a hundred kinds of emptiness we don't even have words for? How many hours had she spent curled on the floor, weeping aloud for the nameless, pervading ache blowing through with the Pacific air? It wasn't so much that the townspeople didn't care about her suffering. It's that they didn't know. They had no idea, just as she had no idea about theirs. I often wonder how many of my own neighbours at any given moment are down on their knees, alone, pleading for their own deep notions of home, with the rest of us utterly oblivious.

The sentiment these days is that nature is our home; nature embraces us, refreshes us, renews us. Returning to nature should feel like coming home. The culture prescribes a regular dose of communing with (sanitized) nature in parks, on beaches, in "adventure travel." It's a kind of math: walk in the woods + reconnection with spirit = batteries recharged for another six months of work (which, of course, is the unspoken point of the exercise).

When she had moved to Ukee six years ago Sara was of that school. But she found that, out here in the midst of all this nature, it's harder to make time for it than it is back in the city. It's sitting right here, but there's so darn much of it. What's special, what's to cherish? What's to make a special trip to see? The beaches are a good walk, but they're just a walk. The rainforest is certainly peaceful, but that's just lack of noise. The ocean is vast, but it's also empty. The storms are entertaining, but eventually it's time to come in out of the rain.

No, nature will not relieve Sara's gnawing rootlessness, her

Low tide at the Tofino Mudflats Wildlife Management Area — an important habitat for migratory birds and many other species. (PHOTO: JEN PUKONEN)

Ancient western red cedar. (PHOTO: JEN PUKONEN)

sense of — let's say it — homelessness. The roof and walls of home are designed to *protect* you from nature, and homelessness is a sore point with her. This came about after the big breakup, about a year ago. It happened in early fall. He found a letter, an innocent request from some researcher to participate in a follow-up study. It was too much for him, and he left town for good, all in about as much time as it takes to say it. Even though the split had been percolating for a long time, it dropped Sara into the limbo of an invisible person finding a place to live.

What single woman can buy a house in this town? She couldn't even afford most of the rents. The apartments didn't seem to be freeing up as they were supposed to after Labour Day, and with no accommodating friends or buckets of money, she looked to the open arms of nature. Believe me, the woods take on a sinister aspect when you face living in them without roof, shower, kitchen or bathroom, and with winter coming on. It was a period of frantic search-and-worry best forgotten, before she finally fetched up in a cold, damp, low-ceilinged basement apartment with no furniture. Exactly the position she was in when she started university. Back then it had been a lark; thirty years later, was this progress? At first she thought it couldn't be, but after shaking the rattles for a few months, she began to see it as a kind of backhanded gift. She started seeing the apartment as a stage set for transformation, and herself as a woman ripe for it. A woman just forty kilometres between this place and a possible new home, a new beginning, just a half-hour drive away. She could walk it if she had to. Close enough to be just down the road, far enough to be unreachable. That's how it's always been for the likes of Sara.

Let me tell you about transformation. When Sara was a young man — yes, a man — twenty-eight years before, she'd had that same sense of disjointed invisibility. Back then she called it "transgender dissonance," a phrase she picked up from a series of psychological referrals, the point of which was to convince these earnest counsellors that sex reassignment surgery would make her *feel at home* in the world. The actual moment of transformation came when she looked in the mirror one morning, about five months after she started on the estrogen, and saw, for the first time, her own breasts, still tiny, but unmistakably there. Suddenly she realized there would be no hiding the transition. She also realized she no longer wanted to hide it, and threw all her men's clothing down the garbage disposal chute. She had believed it would be the fulfilment of a lifetime of inchoate longing, and it was. It was also the beginning of a whole new rainbow of longings: for concealment, for revelation, for companionship, and now, for visibility.

Flying home to Ucluelet from Vancouver, at 3,000 feet she saw what an insignificant speck she was within the landscape, besieged on every quarter by soulless trees and malevolent mountains and an endless expanse of cold water. The only way in or out was by machine, and it seemed that all of them, Tofitians and Uclutians alike, were poised out here on the slippery brink of oblivion.

The thought had never bothered her before, but now it settled in the back of her mind like an odious squatter. Every glance at those looming trees, even the ones in the yard next door, every ripple on that hypothermic water, even in the dead-calm harbour, reminded her of her tenuous predicament.

She realized, viscerally, that the celebrated gulf between Tofino and Ucluelet was nothing more than provincial, small-town thinking. She realized that if Highway 4 were to go, if it slumped and melted in the Big One, if it floated away in a tsunami, if it finally disintegrated for good under the onslaught of all those trucks and RVs — then she was screwed. They were all screwed. That was the secret message of flight 103: *the two towns are exactly the same.* It was the space in between that mattered. There was no separate, improved life waiting in Tofino.

You think of the pioneers coming out here a hundred years ago by boat and you've gotta just shake your head. What were these people thinking? How desperate could they *be* for a place of their own? You're dumped on the shoreline with a few bags of possessions in the midst of all this chaos . . . crazy! You and your cohorts can only go one of three ways: you bond with each other and circle the wagons in a mutual struggle against all that urgent wilderness; you lock down your horror deep in your heart, build a house and never leave it, and see your neighbours as deadly foes in a merciless fight for survival; or you go crazy like the birds, and blend biblically into the wilderness. As the Bible enjoins, *Look at the birds of the air, for they neither sow nor reap nor gather into barns; yet your heavenly Father feeds them,* etcetera, etcetera.

Sara started sorting her stuff into boxes. Weeding out, cutting down. Most of it she gave away. The stuff that would burn, old papers and things, she took down to Big Beach one windless weekday evening, when no one else would be there. She crumpled up some old letters as tinder and laid on some sticks; the pile burst into flame almost before she lit the match, so ripe was it for combustion. She pitched her old papers handful by handful into the fire. She had her hardbound journal, too, with entries dating from last winter. It had some page corners turned down and she read, by firelight, some of the very few optimistic passages she had written back then. You dream it continuously, it becomes your monologue: *She gets unstuck, or comes unstuck, or unsticks herself, or meets someone who helps her come unstuck, and she starts to move. Or feels she is moving. With motion comes direction; with direction, purpose. Things fall into place: a house, a companion, peace of mind. Eventually she realizes she was composed for ecstasy, primed for it, that she'd lost the path but has now found it again.*

Into the blaze it goes.

So here's what's up with Sara's future: She does move, but not to Tofino. Or not *just* to Tofino. She moves to the Pacific Ocean. She moves to planet Earth. Money comes to her, she doesn't understand how. The rattles bring it, maybe. With the money she buys a boat. She doesn't know how to operate a boat. But she can learn. She has to learn, because now she lives on a boat. She turns her back on land and its concerns, its empty roadways stocked with invisible wanderers. She doesn't stop worrying about money, of course. Let's say money stops worrying about her. She lives on the water, awkwardly at first, eventually with more grace. She putts back and forth between Barkley and Clayoquot Sounds, choosing her weather with great care. She gets a reputation as a local character. She'll know quite a few boat people soon, but she doesn't usually join them at the docks. Often she anchors out in some deserted cove in one of the sounds; it won't be long before she's got a list of dozens she can pick from. But perhaps her favourite place to be, when the sun goes down, is anchored a short distance off one town or the other, a scant quarter-mile of water that makes all the difference in the world.

What she often does, on evenings when weather permits, is to brew herself a pot of jasmine tea and sit out on deck in a folding chair, watching the world grow dark. The evening's clouds coalesce. The sun gives up its burst of colour. The air goes misty and cool. And the lights of town come on. She'll sit there for an hour sometimes, bundled up, watching, peering through binoculars at car headlights moving along the roads. Not curious, just watching.

Sara finds it marvellous that there are people in those cars, alive, out on definite missions. People full of thoughts and passions, none of which travels across the water. The unmoving watches the mobile, the solitary tracks the social, the invisible spies on the illuminated, from a floating beachhead in no-woman's land. But it's not a war at all; it's a kind of peace, brokered by salt water and the night. Eventually she stands up, stretches, and turns in.

So, When Are You Moving Back?

- JANIS MCDOUGALL -

Heart had a lot to do with it. Brain was definitely not involved. I was twenty-one years old when I was lured to the "wild west" — living in the moment, following the love of my life and trusting my gut. Now with my middle-aged wisdom, I recognize the body's intuition. Whether it comes from your gut or your heart, both are sensitive to the energies around, and acting on impulse is following the heart. Pulse is the indicator of life. When someone or someplace "feels" right or compatible, it's worth paying close attention, no matter what others may say. Brain, be quiet.

Some of my urban friends thought I was either uncharacteristically rebellious or dramatically romantic to be living at "Long Beach," as they called it. They thought it was just temporary and wondered when I would move back. How could I convince them that there was no need? I had everything I wanted within reach, as well as a deep sense that this was the place to be.

For two years I lived and worked on Meares Island, right in the belly of Clayoquot Sound, long before the media exposed "the Clayoquot." In the 1970s, the name was never mentioned in conversation, or used to refer to the great expanse of wilderness and waters that stretched from Kennedy Lake to Hesquiat. There are days I wish I could go back to that time. The time before the world discovered Clayoquot Sound, when no schools of kayaks cluttered the harbour. No clutches of charter boats "hunted" for whales. Espresso and cappuccino were foreign words never tasted. The rustic Schooner Restaurant served monster burgers and fries. Locals hung out there, wearing grey Stanfields, blue jeans and boots.

At that time, the handful of surfers in tattered wetsuits openly cut through vacant private properties to check out their favourite breaks. At one popular resort, surfers were invited to create a separate lot to park their old rusty wheels. The owner believed that the sight of shaggy surfers sharing the beach would entertain his guests. It was easy to recognize those who worked at the fish plant by the flash of silver scales on their clothes and by the invisible cloud of odour that remained in the bank long after paycheques had been cashed. Those were intimate times. Fewer crowds, more space, shacks in the bush, empty beaches, plenty of fish and recognizable faces.

My first home in Clayoquot Sound was a little shack right at the base of the immense volcanic-shaped, evergreen Lone Cone. In my humble shelter I could almost wash dishes in my sleep, as my bed was in the kitchen. The window over the sink faced west, and there wasn't a hydro-powered light in sight. Vargas Island broke the horizon. Tofino was not in view.

I learned the tides of patience compulsory for dwelling on a small island. Living in the moment had to "wait" for a while. Frequently, involuntarily, I practised waiting. With longing, I would watch for a boat to come, hoping for a visitor. I was always scanning the horizon. I could quickly distinguish the shape and size of a vessel the same way one recognizes the familiar posture and gait of a loved one. Looking beyond my steamy window in the dark, I anticipated the delight of a slow floating light, with a distinct size and brightness. It signalled I would have good company. I waited for arrival. I waited for departure.

Many times, with resignation, I stood on the dock after summoning a water-taxi to take me away for a temporary break in the village of Tofino. Depending on the driver's activities or whereabouts (Ahousat to Opitsat), I could wait from fifteen minutes to an hour. The local water-taxi *Silver Stud* cost five dollars from Tofino to Kakawis or vice versa. The actual trip took about five minutes. The ramp between loading dock and wharf varied in degrees of steepness depending on the tide and the swell. During some raging storms, walking up or down the precipitous rolling ramp was like riding a mechanical bucking bull. I was at the mercy of the elements but I never thought about quitting or moving back to the city.

There was something about the yin and yang of adrenaline rushes and periods of patience that made me feel balanced. Strange noises in the night could easily be identified. The first time I heard the "mad" bovine, I thought it was in my room with me. The wild cows of Meares Island took relief at rubbing their sides on the siding, and the boardwalk ran beside my place. The cows had no toes to tiptoe on and they rubbed and clomped at the first sign of dawn. It was comforting sharing my space with these creatures; that gut-heart connection relaxed with an awareness of organic energy. I easily fell back to sleep feeling safe.

Occasionally, of course, I went to the city — more to see loved ones than the bright lights. Heading back to Tofino after these visits was quite the long journey but never one travelled with dread or regret. After passing the West Bay and speeding by Sproat Lake my excitement would grow with each bend in the road. I basked in satisfaction with the return to home. I was young. I was bold. I was motivated by love. The quiet, gentle surfer who enticed me out west was crazy enough to settle me down. My cabin at Kakawis just wasn't close enough to his home in Tofino so we set out to create a nest of our own. The first step was buying some land.

No research was done. There wasn't even a real estate office in

Tofino. Small "For Sale" signs directed us to what was available. The village clustered around the conveniences of town, the waterfront, the hospital and the school. We went searching outside the town limits, which ended at the trailer park by the only gas station. We preferred some undeveloped land. The lots across the gravel road from Chesterman Beach just seemed too swampy, unprotected, and open to the neighbours. The natural vegetation appeared sparse and stunted. We wanted large and lush, not a quarter acre lot on a bog. Those calculating with their brains would be laughing at me now, I have never been much of an investor.

But I listened to my heart and the unspoken communication that can only be described as the spirit of place. My partner had a friend with partially cleared land that was too far from town for his family. We both had good feelings the first time we saw it. I remember the calm that grounded my body when I stood on the secluded lot: cedar, hemlock and salal all around. The bush and the trees sweetened the air and offered a welcome-home blanket. This was a place for the living.

I considered it "no-man's land," an area of undiscovered beauty, the perfect place for peace and privacy. It was halfway between the centre of town and the south end of Chesterman Beach, which was about as far as people in the community were officially living, back then. We settled up with the bank, then we settled down. The house was designed with practical features. Many salvaged materials were carefully recycled, purchased after scrutinizing the *Buy, Sell, and Trade.* The Buckles of Vargas Island provided freshly milled lumber. Fill for the septic field came from Bob's mill. Water was located with the help of a diviner, my first exposure to "energy work." What made him divine was his mysterious talent, and what surprised me much more was his nondescript appearance.

I was somewhat sceptical as he climbed out of his truck in his leather work boots, worn jeans and an open blue-collar shirt. He could have been any other Port Alberni labourer. His hair had the shine of Brylcreem and his hands were rough with calluses. His only tool, the forked tree branch, no longer than a metre, was disappointingly average, but what he did with the stick made the difference. It was claimed that he could source an underground natural flow of water, though he wasn't one to spout off. His divining rod was held parallel to the ground in line with his sensitive heart. With a forked end in each hand, he roamed the property in search of an invisible force. As if waltzing, he gracefully guided the branch, giving his dancing partner his full attention. All of a sudden, the tip bent to the earth as if pulled by a strong current. "Y" marked the spot for drilling.

Water pumped pure, although slow, from the well as we turned to construction. The roof went up in an October work bee. My job was feeding the volunteer crew, but I got my share of nail pounding. The air filled with the smell of fresh cedar shakes, noisily split right on the scene. Chatter and laughter eased the tedious work and combined with the percussion of hammers. The whole roof was done in one day. More than twenty years later, when the roof had to be replaced, a rusty souvenir hammer was delightfully discovered up on the rooftop under a shake. The purpose of having a roof over our heads became more than protection from the weather, it was part of the labour of nesting, and it was successful. We welcomed the amazing arrival of a child.

Raising a daughter gave my life new reason. Invisible roots were planted. I was on the branch of the ancient family tree belonging to my only offspring. I marvelled at her transformations as she grew from clinging baby to confident young woman. Growth was not hers alone. She blossomed while Clayoquot Sound boomed, as we sought comfort in the buffer of our home. Now she temporarily lives in a city and when she meets new people she explains that she was born and raised in Tofino, and that sense of place, of home, becomes an icebreaker, drawing curious people even closer. And they inevitably ask her, "So, when are you moving back?"

I still get asked whether I would not prefer living in the city. There are times I admit I am tempted. I have had my share of crises and there is always the mystery: would it be easier if I left? Easy is not necessarily better. What more does the city have to offer? Three months of each year, I don't have to wonder — the city comes to Tofino in the form of crowded, busy summers.

I live on my own now in the house built with hope. It stands on the land chosen a quarter of a century ago. From every window I see shades of green. All shapes and textures provide natural inspiration. Solace whispers like a breeze in the leaves. Huge trees stand guard outside, sheltering and attracting wildlife. Great blue herons, waiting out wild winter storms, have clung to the candelabra cedar. Spring's blossoming salmonberry invites wee rufous hummingbirds. Ospreys have cried while gliding overhead. Raccoons prowl the yard when it's dark.

The innocent salal lured the biggest surprise, reminiscent of waking to Meares Island cows. I remember waking before dawn at a time when salal berries were plump on the bush. My window was open but covered with a screen and the sound of leaves rustling disturbed my dreams. I could have ignored that but for a deep husky coughing: a feasting black bear choking on stems. The bear soon returned to its bittersweet treats and I laughed myself back to sleep.

All of the beauty, the wild and the space would never be enough to survive on. Nature never nurtured all of my needs. I complement my life with community. I keep in close touch with my family. I am blessed with unconditional friends. Home is much more than a building; it's the place where I feel most secure. I can easily say what brought me here. It's difficult to say why I've stayed. In short, I did it for love. Over time, I've conditioned my heart not to depend on another. My heart now belongs to a place.

The Sensual Coast

Living in the Everyday

- ANITA SINNER -

What is important about living here? What can I share of this life? How do I encapsulate my sense of this place I call home? For sixteen years, I have lived on the southern tip of Vancouver Island in the community of East Sooke, and today I write about the sensuality of living coastal, about the intimacy of the everyday, in between the contours of the natural world, of the land, the sea, the sky. My questions are openings to deeper reflection, to a meditative state with nature, where I draw attention to what is exceptional in the mundane, and from which I seek to construct meaning. I am inspired by the cacophony of song that now surrounds me, an operatic performance of countless birds hidden in the stillness of the thick forest. I write of finding happiness in finding home. These moments enable me to live creatively, aesthetically, artistically, *coastally*. This is my love of place, my topophilia, my west coast.

The experience of living coastal may best be articulated in a series of interludes, short vignettes inspired by my passion for home. These ruminations are glimpses into my experiences of living aesthetically, as I explore the interrelationship of nature as sanctuary and my source of spiritual renewal in a landscape where I seek to compose a life in a coastal realm.

And so I begin.

INTERLUDE: THE PRIVILEGE OF SOLITUDE

We came to our cottage in the late 1980s. It was a daring venture for by-products of the city, moving from the then most densely populated community in Canada, Vancouver's West End, to an outlying community of just a few hundred people, with the nearest town just a few thousand, on the bottom of the big island. To come to a landscape and find home, leaving behind the understood, and discovering what place can mean, is a wonderfully romantic, back-to-the-land perspective. It was not always easy, but it has been a source of great joy. Finding our way here was quite unexpected, and yet my partner and I have since rooted in this ground, our postage-stamp third of an acre, in ways we still do not fully grasp. We remain, year after year, and I cannot foresee leaving this place. I have become part of the permanency of the land, in the details, for here I find peace, in my hermitage, I am me.

I look across the backyard now, a meadow of hundreds of English daisies, with only splashes of grass. Here I have often studied the intricate passing of shadow and light in subtle changes of colour, texture and patterns. I feel the harmony of the land physically and emotionally, with a spiritual quality of being that is my own. Living coastally is a site of seeing, a sensory and perceptual way of knowing, where the imprint of the land upon the soul is a lasting signature.

INTERLUDE: GREED BREEDS MEAN DEEDS

I noticed this simple phrase in passing, sprawled on a cenotaph in Sydney, Australia, a piece of graffiti that remains with me to this day. The greed that breeds mean deeds in my coastal experience is money for trees. Despite the active role of the environmental movement to save tracts of public lands, the private land question remains unchallenged. Even though the care and cultivation of vegetation on private lands is a legal and communal requirement in many countries of the world, we do not have that custodial tradition or even simple respect in British Columbia. It has been a free-for-all, where greed breeds mean deeds against nature.

The transformation of coastal landscapes in the last decade has been traumatic to those who live aesthetically. As large acreages along the coast are sold, they inevitably become clear-cuts. Given the price of land, buyers are often not local residents, and the resulting actions of buyers are defended as "good business" decisions, which are made easier because they do not have to look at their destruction. They are absent. The value of magnificent trees is just too tempting. Section after section, everything is eaten by monstrous machines churning, cutting, trimming and bucking tree after tree. So many properties denuded and left for the next buyer. The only element slowing the complete destruction: the price of trees has dropped for a time. It is only a reprieve. It will not last.

For seven days
a caravan continues,
rumbling past my door:
Logging trucks loaded
with hundreds upon hundreds
of spruce and cedar and fir.

So fresh is their cull
that the blood-scent
floats sickly sweet
and lingers in the air.

As these rolling coffins of steel pass,
the standing trees seem to call
after their murdered kin,

in a sudden gust
of fear and grief
that stirs between their branches.

INTERLUDE: HANDS PASSING OVER THE EARTH

Come late spring, when I spend my days outdoors, it is only when the westerly winds pick up in late afternoon and the laundry dances on the line that I am reminded the day is passing. Despite a life of learning the social necessity of following structured time, of watching the clock, I rarely seem to know what time it is. In our home, all clocks show different times, and not one is quite accurate. This was not a purposeful act, but a slow evolution of living coastally. We are subject to numerous power outages, and the task of resetting clocks is frequent, very frequent, and over the years, we reset fewer and fewer, and eventually, we have a house where time is, in our own way, inverted.

One of my favourite experiences living here happens when the power goes out. For a brief respite, we exist in absolute silence. All is still. Without power, we recapture time in a curious, almost historical way. I regard this quiet serenity as a precious gift, often a gift of nature during a storm, impeding the taken-for-granted usage that we depend upon. Only when the power is off do we remember what silence really means. Only when the power returns, do we realize the immense loudness that surrounds and fills every waking moment. We have grown too accustomed to the constant hums that fill homes. We forget the influence of electromagnetic fields that pass through us during the course of the day, the many sources of noise making, and the weighted tension generated by television, computers, everyday machines, chaotic movement. Is this why we treasure our time in nature, to escape the drone of electricity, and the endless burring of the fridge at night? Is there something in the post-modern human spirit that craves a return to a state of silence?

I suppose I am among the tribe that sits by the window, absorbed in the mundane task of waiting, suspended, attending to a falling leaf, enticed by the transformative shift of this leaf meeting the forest floor. Another day begins with a sense of nurturing wonder. Living coastal is a delicate state that encourages sensuality in such moments, suspended in time, making space for imagining and contemplating.

INTERLUDE: AN ARTIST'S PLACE, AMONG MANY

Why do we decide to "go coastal?" How does this decision ultimately remake the life path? I wonder about our relationships with nature, as well as the geological underpinnings of living on the coast. If we consider the energy of the rocks that surround us, the rocks upon which we build our homes and live day-to-day, could the magnetism, the composite make-up, alter our moods and perhaps even influence our chemistry? Could this sway our wants and wishes and dreams? Is not geology also part of the interrelationship of self in the natural world?

There is something different about this place called East Sooke, a small peninsula composed of lavas and igneous rock, the signs of very ancient volcanic activity on an island arc.[1] These rocks still tremble, now in response to the movement of continental plates along the Juan de Fuca ridge, rattling our windows and glass doors in the seemingly ever-present clustering of tremors that remind us "the big one" has long been promised.[2] On occasion, when the earth begins to roll for brief seconds I wonder, is this it? With just enough time to comprehend the possibility, though not enough time to respond, another small quake passes and we carry on, unstopped by recurrent erratic messages that the earth beneath our feet is never constant despite the solid rocks that form the outer crust.

But could these rocks also influence our aesthetic choices, our artistic expression? As I come to know more residents, I discover more and more creativity. It seems that in almost every household an active artist resides: painter, writer, potter, musician, wood turner, performer, photographer, crafter and more. Is there some intuitive response to the energy of the rocks and the seismic trembling that draws artists to this place, making home a sacred place, a place where we feel so deeply tied to the earth that we are within the very seams and layers of the geographical formations in our midst?

Perhaps in part, because of dynamic geological events, we manifest ourselves through the arts, while in calmer landscapes creativity may not come to fruition in similar ways. Are those of us who find we cannot live elsewhere more closely tied to these tensions in the earth? If the integration of that tension is foundational to the creator-artist, then it is possible, when a landscape is geologically felt, a desire for expression awakens. Of course, I recognize this is a rather unorthodox interpretation of events and people in their quotidian lives. Yet it seems unusual that in a small population there is such a high concentration of creativity. The intense rhythms of artistic practice, inlaid with the pulsating forces of the most seismically active region of our country, reflects a generative space where cultural knowledge, traced in coastal living, is collectively formed through a vast array of artful expressions. Maybe in the geological margins we find a different kind of freedom that enables the spirit to express something original in a world increasingly consumed by consumption. Maybe that is why we come to, and stay with, these ancient rocks.

INTERLUDE: A PARTING NOTE FROM THE SENSUAL COAST

I lay down my pen. The late afternoon is so warm, the day so peaceful. What a privilege this is, to live coastally. I watch a pileated woodpecker scale the standing dead tree trunk that looms above our cottage. The routine is the same: the woodpecker surveys the advanced decay to determine where grubs are most active, then he peels the bark from the rotten core, and the drumming begins, leaving a mass of wood fragments scattered below for me to collect. And once successful in locating insects, a joyous call synonymous with Woody Woodpecker rings out, and just as quickly is answered by a mate that is never far away. In our time here, only one pair has consistently made the surrounding forest home and I wonder if this is the same pair of elusive birds all these years. After months of this practice, the tree has begun to lean, and will no doubt fall back to the ground before long. I consider removal, since insurance demands the interruption of nature's course in such matters. But I prefer to let it be.

Perhaps sharing these everyday moments in the act of writing is too idyllic a portrayal, but this is what reverberates most now, in this moment. I omit the darkness of local trauma, violence and crime that in some ways seems much more severe when it occurs in a small community. Those stories belong to another forum, another day. Instead, I strive to highlight the interplay of self and nature as a process of being attuned to sensual qualities. I believe that in the fluidity of every moment, and in the resonance of every form we behold, something transformative happens that defines these coastal places we call home, innately, subconsciously, permanently, embodied in heartfelt knowing that can never be fully articulated, waiting for us to stop, to see, to feel, to stay. Now it is time for a walk by the seashore, to meditate with the rolling waves that flow over and back. I follow along in anticipation. The pebbles pop and bubble under the tickle of the water's touch, and like shimmering sequins, sparkle with the last light of the west coast sun.

REFERENCES

Strahler A., & A. Strahler. (1983). *Modern Physical Geography 2nd Edition.* Toronto: John Wiley & Sons.

Yorath, C., & H. Nasmith. (1995). *The Geology of Southern Vancouver Island: A Field Guide.* Victoria: Orca Book Publishers.

NOTES

[1] See Yorath & Nasmith, p. 17.

[2] See Yorath & Nasmith, p. 53 and Strahler & Strahler (2nd Edition), p. 216.

Encountering

Sex in the City

Love in the Forest

- BRIONY PENN -

Twenty years ago, *the* place to kiss amongst my cohort of twenty-somethings was any mossy rainforest on the west coast of Vancouver Island: Clayoquot, Kyuquot, Nootka, Barkley — any of those sexy-sounding sounds — was the place to seduce with beauty, fresh air and sparkling salt spray on your skin. There was no shortage of secluded fairy glens among ancient trees to lure your sweetheart. In the perfect comfort of a mossy bower festooned with witch's hair and glistening fern fronds, hearts were awakened and you emerged the naturalist/activist dedicated to saving these sensuous places — gods and goddesses amongst the lichens and liverworts. For our efforts, we ended up with a good knowledge of the flora and fauna of the temperate rainforest, the prison sentences and/or the progeny of our love. Twenty years later, as a mother of

Reprinted by permission of the author. Originally published in Monday Magazine.

two, bespectacled in a stuffy university office (with only a hint of sea spray on my lined cheeks), marking essays on natural history by twenty-somethings, I realize things have really changed. The underlying sub-themes of essays used to be love in the forest; now it's sex in the city. This all poses a problem, because the last ancient rainforests on the island that we thought were protected are going under the axe and the goddesses needed to protect them have been seduced by the city. I'm also not convinced that they're so happy in the city either.

I know the goddesses in the city are suffering because I've been doing some research. It all started when I realized, as a teacher of natural history, that there was a real shift in the student zeitgeist. There seemed to be a better understanding of sex toys in New York City than the liverworts in Clayoquot Sound. Teaching liverworts to someone when there is no point of reference or connection is difficult. So a friend with a twenty-year-old daughter lent me a video of the TV series of *Sex and the City*, and I sat down and watched an episode. It involved endless rapid couplings with business CEOs in different New York apartments filled with consumer goods. Compared to walking on a golden sand beach in Clayoquot with a full moon catching the white-crested waves beside a hunky tree planter quoting Hafiz, the video left me with an extraordinary sadness. Is *Sex and the City* the benchmark we have set for the place of choice for mystical unions? Is this what the next generation will go to jail to defend? I would go to jail to prevent a life spent like that. I couldn't even tell them apart — the executives or the apartments.

So if the city has supplanted the rainforest as the place of desire, the place of love where one forges a cultural and spiritual union, what are the repercussions? I think it is enormous when you realize that it was the small intimate rituals that we experienced with plants and animals that rooted us and connected us to those places. When I see a salmonberry patch, with a Swainson's thrush singing its spiralling love call, I think of early summer wanders with berries dripping from our lips and the first hot rays of the sun burning through the mist into our damp clothing. It was a time of perfec-

Meares Island, Clayoqout Sound. (PHOTO: JEN PUKONEN)

Ladybug on silverweed. (PHOTO: JEN PUKONEN)

tion, a benchmark against which I measured my life. If I was to have formed a loving bond in a brand-name bed with designer sheets in a multinational hotel room, then it is that place that my memory would have cast back to, not the salmonberry patch.

But to answer the question of whether the city has become the place of choice, I decided to find out what the quintessential sex-and-the-city women really considered were sensual places. I went online and talked to Ana. Ana is one of the original web-cam girls, featured in the documentary *Web-Cam Girls*. Ana makes a living as a sex goddess/installation artist of the city, training a video on herself twenty-four hours a day within her four bleak walls. She has been doing it for nearly a decade. You can log on (for a fee), watch constantly changing still images of her and chat with her on line. When we logged on at the premiere of the film, she was sitting alone with breasts bared and a painted sad clown face amidst several cakes with candles lit for the occasion. I asked her if she had ever thought about wandering in a rainforest on the west coast of B.C. She typed back right away, "Could someone send me the fare?"

Now the only temperate rainforest a web-cam girl in New York is likely to have seen is Clayoquot Sound when CNN was broadcasting the goddesses and grannies going to jail. Ana had also seen Nootka Sound where the "lonely" whale lived. "Could I see that lonely whale, I've always wanted to see a whale?" she asked. Think of it, wild places on Vancouver Island reside in the minds of city sex goddesses as a place of sanctuary and sensuality. I didn't have the heart to write and tell Ana that she had better come fast. There's not much left of Nootka Sound and they are going to be logging seven valleys of Clayoquot Sound. One of those valleys is called Pretty Girl Valley. Ana is a pretty girl, but she would be beautiful striding free through that forest with no video trained on her cleavage. Another valley is called Tranquil Valley. We all need tranquility at some stage in our lives. Who knows if those two valleys will even still exist as they were when Ana makes it for a visit. If they don't, Tranquil and Pretty Girl might be just place names with no vestige of their namesakes. Ana will not enjoy them. Nowhere in the literature of the world is a clear-cut a place of sanctuary and sensuality.

So what is happening to the zeitgeist of the young? Twenty years ago, there would have been lots of young people with experiences derived from robust childhoods and summer jobs exploring the west coast. When children wander whimsically in the intertidal zone or gaze at a star-studded sky from a mossy bower, they come to know the sea stars and their celestial counterparts. Ecological knowledge is disappearing so fast that it is more endangered than the Vancouver Island marmots, because environmental education has disappeared. Go to any biology department of a university and you have to look hard to find the ecology section. They are the unfunded, poor cousins stomped into oblivion by the biotech industry and the professional vandals of the corporate university. Opportunities for children to connect with nature have diminished to maybe a couple of encounters in urban parks heavily supervised by fearful parents, a harried program on earthworms or a rapid whale-watching tour. Go to any museum, school or park and naturalists have been dropped from every program.

My university classes are full of young adults who have never held a shore crab in their palm or spent a night in the wild. They haven't a clue where their water comes from or where the moon sets. I see them sending notes to each other electronically and long to tell them about the beauty of a lover's whisper in a silent rainforest where you can hear each other's heartbeats pulsing in time with the waves lapping at the shore. When I travel up and down the coast and look at who spends time outside observing the daily comings and goings, it isn't young people. It is mostly an older generation. There are few jobs anymore demanding local knowledge, except the rare ecotour guide. We don't even have young loggers or fishers anymore with bush and sea knowledge. They know more about "logging on" than logging and are heading to the city, thinking that's where life is at. This is the irony of having taken so much so quickly. If we had developed a stronger sense of place fifty years ago, we would have lots of young people who would have been taught enough about the forest that they could understand the logic of retaining these places for both mystical unions and physical sustenance.

But my conversation with Ana has left me realizing that there are more people out there who care for our forests than it would appear. What is stopping many from speaking out is that they don't know anything about these places and fear to appear ignorant. But they are moved by basic impulses of beauty and something higher than a boring apartment filled with consumer goods and a mindless relationship with a corporate soul who probably longs for an honest day's sweat in his life. I keep thinking about Ana and the other young women and the CEOs down there managing their forest industry portfolios between couplings and wish I could just bring the whole damn lot of them up to Pretty Girl Valley and Tranquil and let them experience beauty and integrity for once. Maybe they would see that love in the forest beats sex in the city hands-down any day, and that their deals to liquidate the forest are in fact deals to liquidate themselves.

On a Quest for the Western Screech Owl

– CAROLYN REDL –

Think Tofino, Opitsat, Ahousat, or any other little village up the coastline of Clayoquot Sound and you're likely to visualize those bright orange survival suits bobbing up and down: whale-watchers sitting side by side in zodiacs, gliding along on the sweet swell of the Pacific, all eyes peeled on the horizon, watching for grey whales to slip beneath the surface with a stupendous flash of the tail. You might see those things the tourist brochures list: sunsets over the open ocean, treks on boardwalks through rainforest, rhythmic clanging of buoys, and steamy soaks at Hot Spring Cove. Not I. What I see when I hear the words "Clayoquot Sound" is a little innocent bird, the western screech owl.

It all began one afternoon when I came home from work to find my partner, Hans, standing beside the van, in which a pile of sleeping bags lay atop thick foamies and smooth Thermorests, all under a few blue plastic crates, bulging with food and miscellaneous gear.

"Got your suitcase packed?" he queried, knowing full well this was the first hint for me of a trip. "We're going to the west coast to find the western screech owl," he announced, as casually as he'd tell me that we were meeting his mother at the airport.

I threw a few things into my bag — gumboots, heavy sweater, toothbrush, sweats, trusty Gor-tex overpants and jacket, the things I'd likely need for at least some time on a February weekend, given Tofino's annual three and a half metres of rain.

Soon, we were chugging over the hump to Port Alberni. The engine hummed smoothly and we sang along with the voice on our sole CD, the one we've been meaning for months to exchange for more recent releases. We arrived in Tofino and had a quick and tasty fish and chips. Then, because there didn't seem to be anything else to do at that time of year on a drizzly night, no movies or community get-togethers, and we didn't feel like hanging out in the smoky bar, we drove down to the surfer's beach. In truth, Hans kept talking on and on about the western screech owl, only eight or nine inches tall, weighing between five and nine ounces, and recognizable by its distinct ear tufts, facial disk and black semi-sideburns.

"Did you know that its Latin name is *Otus kennicottii, Otus,* meaning 'horned,' yet the structure is not a true horn? The species name, kennicottii, is the Latinized name of an American naturalist, Robert Kennicott."

Whenever Hans starts defining Latin names, I know he's exhausted his knowledge on a subject. At any rate, by this time, we'd parked beside a surfer's van and needed to consider the practicalities of camping. The government campgrounds were all closed for the winter. I rolled down my window.

"Can we stay here overnight?" I asked one of the fellows standing beside a Coleman stove perched on a log.

"Nice night," he replied. It seemed awfully wet to me.

"Good surfing today?"

"Not so bad." With him, I reckoned, I wouldn't make any headway addressing the subject head on.

Hans leaned across and asked, "See any screech owls around here lately?"

"Oh, you guys are some kind of birders, are ya? There's been a pack of wolves seen regularly round here, but, na, haven't seen any screech owls. Matter of fact, haven't seen any kind of owls. Couldn't hear any either, what with the surf." He gave a wave oceanward, turning our attention momentarily to the deep, rumbling breakers thundering landward, not a dozen metres from the rim of driftwood logs separating us from the spray.

"They're pretty common, though. Heard there's been some up at Kennedy Lake."

"Yeah?" Hans practically leapt onto my lap to hear better. "Whereabouts?"

He gave Hans directions to a road past the landfill. "Straight ahead, can't miss it," he said.

"Maybe we could just go to Victoria, Hans," I interjected, "where the western screech owl was recorded on the Christmas bird count as the most frequently seen *and* heard owl. We wouldn't have any trouble finding one there."

"You could camp here tonight," I heard the guy say. "The cops come by now and then, but just to check there's no trouble."

No point pushing the Victoria idea, I decided, as Hans turned off the engine. The stranger went back to his meal, and we began arranging stuff in the back of the van so we'd have space to sleep. We rolled down another window a tad, crawled between the layers, and listened.

"Rrrrrrpp." Pause. "Rrrrrrpp." Pause. "Rrrrpp."

My heartbeat slackened and fell into rhythm with the surf. The steadiness, the insistent regularity was heavenly, and it seemed almost the next moment I was opening my eyes. It was light already and surf sounds had changed to a muted rumble. I looked out to see heavy mist and waves breaking far beyond the shining smooth sandy beach exposed by the outgoing tide, and already the surfers were out playing in the breakers.

Given that owlers get best results at night, we spent the day wandering up and down the coastline between Ucluelet and Tofino, exploring side trails and tide pools. We ran from one end of the bay to the other at Long Beach. After supper, we lingered in the restaurant to watch the crashing waves until we could not see anything.

"You don't really expect to drive on that logging road and, just like that, find an owl, do you?" I asked.

"Of course," said Hans confidently.

I suppose I should tell you a little bit about Hans. He's an avid birder. That's "avid" with a capital "A," and, further, if he gets an idea in his head, he won't let it go. After six months without success, he still regularly drives his designated route for the Owl Watch Program. Stops every kilometre and listens for owls. No luck. I quit going with him after the first try.

"Tell you what," he added, "if we don't see an owl, we'll come back here and sleep in this fine hotel." He extended his hand across the table. "Scout's honour." My sweetie is such a darling.

At least the rain had stopped by the time we made the turn onto the logging road that forked off past the landfill and continued on into the forest. Probably a dozen times, every kilometre or so, Hans stopped the van, rolled the windows down, listened carefully, and then, with only the silence of the night answering our practised calls, we rolled the windows up again and drove on. By this time, it was very dark. If the full moon was up, it was well hidden behind heavy clouds. The high-beam lights of our van showed trunks of young cedars, hemlock and thick salal bordering the gravelled road on either side. Suddenly, we were on the Kennedy Lake Bridge, the same bridge where, in 1993, protestors stopped trucks, blockaded the road and gave notice to the big logging companies set on clear-cutting all the old-growth forest indiscriminately. Those protestors were ordinary folk for the most part, and they were arrested for their fight against the annihilation of ancient rainforests. We stopped, rolled down the windows and hooted.

"All I hear is the echo of the protestors' chants and songs," I told Hans.

"Funny. Funny. We'll try a little further along, at the edge of the lake bush." We came to another fork in the road. I was beginning to wonder where exactly we might be.

Hans braked. "Look, Carol. Oh, my God, look at that."

There in our headlights, perched on a bare branch, sat the object of our quest. A western screech owl.

"I can't believe it. It's not real. You've been out here and planted a plastic owl! It's one of those phoney owls people put up to scare gulls."

"It's real, an absolutely genuine western screech owl. Sh-sh. Oh, this is too much," he exclaimed. "Be quiet now. I'm going to roll the window down, really slow."

The bird sat there, wide-eyed and stock-still. You could see the yellow of the eyes and the dark pupils. The bird was staring me straight in the eye. Eyeball to eyeball, or, so it seemed. It blinked, but showed no signs of fear. It was like an actor on stage, highlighted square in the middle of our high beam. It fluffed its feathers like an actor might shake his shoulders slightly to settle the costume properly on his frame when getting set to perform. I finally realized that I'd stopped breathing and took a slow, shallow gulp of air. This was mesmerizing. It was hypnotic. I did not want to move a muscle. What did it take us for? Prey? Just mere light, coming out of nowhere, like a ground-bound mini-sun?

"I'm going to turn off the engine and see if it will hoot," Hans whispered.

"Don't scare it," I answered, as if I were its protective mother.

As Hans turned the key off in the ignition, I took his hand and held it to hold him in place. As the engine died, the bird suddenly darted to the ground but we could still see it in the headlights. It caught something and, to my surprise, flew back to its perch. Its feet grasped the branch. I think it was an alder branch. The owl slowly settled its feathers. It swallowed. Stared at us and blinked.

"Wha . . . Wha . . . Wha . . . wha . . . wha . . . wha . . . wha . . . wha . . . wha," Hans gave the call, the sounds gradually diminishing in volume, like the bouncing ball noted in all the bird books. I held back my laughter, thinking of the absurdity of it all. No self-respecting owl would fall for that call, especially when the source was immediately in view.

"Wha . . . Wha . . . Wha . . . wha . . . wha . . . wha . . . wha . . . wha . . . wha."

"What?" I felt my mouth drop, "Was that you?"

"Wha . . . Wha . . . Wha . . . wha . . . wha . . . wha . . . wha . . . wha . . . wha," Hans repeated his call.

"Wha . . . Wha . . . Wha . . . wha . . . wha . . . wha . . . wha . . . wha . . . wha," came the reply.

He tried again, but the owl stopped cooperating. It sat there, looking on, giving us time to count the rows of light and dark feather tips. It was overwhelming: the whole idea of a western screech owl sitting for us in the forest. I felt myself drawn into an aura of mystery and magic, of timeless presence without beginnings and endings. Life epitomized in a rolled-up ball of feathers, sitting in watch over and with me, the space between us given over to eons of patterned differences and similarities drawn into one life web. All held as a clasp holds pearls on a thread, tenderly, delicately. The owl darted groundward again and then repositioned himself on the branch.

"Must be some insects out there that appeal to him," I pondered.

"Sh . . . more likely mice. Little mice," Hans replied. "Look. Its tufts move when it blinks."

"Did you bring the camera?"

"Didn't you?"

"No."

"Well. It's too late now anyway. Look, it's gone." While we said those few words, distracted for a few brief seconds, the owl had disappeared as quickly as it had come into our lives.

"Tell you what. Let's sleep in a real bed tonight," said Hans. Miracles would never cease, I thought to myself. First the owl and now this.

"Seeing a screech owl at close quarters is cause for celebration. We'll go back to town, have a late snack, and nestle down in an oceanfront room. We'll have breakfast in bed tomorrow morning.

Maybe it will even storm and we can spend the morning like honeymooners watching the crashing waves." Hans can be a totally incurable romantic.

Never mind. Never mind that, for when Hans turned the key in the ignition, nothing happened, not a spark of life was left in the battery. We had apparently watched the owl much, much longer than was prudently manageable. We had not even noticed the headlights dimming. Hans lifted the hood and jiggled the wires. He got back in, tried again. No response. All this in total silence, too, if you can imagine. Our trusty van was not budging an inch. Well, we weren't destitute. After all, we had our camping gear in the back and plenty of food. We could survive in the bush for several days. Hans wasn't fazed in the least by the latest development and quickly prepared the bedroll. Before I could complain, he was snoring.

Sleep didn't come quite as easily for me. How would I ever explain my absence from work on Monday morning? Even if we'd had a cellphone, we would not have been able to call. We were out of range. And how would I describe the predicament? Even if we got up by eight, which we would not, I knew, we could not possibly hike the seven kilometres back to the highway, get a tow truck, return to our stalled van, and get back home, all in time for me to drive the forty more kilometres to arrive at work by 8:30 A.M. on Monday morning.

"You went where?" I could hear Kathy ask.

"On a search for the western screech owl," I would have replied, meekly.

Can't you just hear my colleagues giving a hoot every time thereafter when they passed me in the hallways, crowded with students, all eager for a little fodder for razzing the teacher?

"Wha . . . Wha . . . Wha . . . wha . . . wha . . . wha . . . wha . . . wha . . . wha."

It would make a great story, at my expense, nonetheless.

I remembered the phrase, "pack of wolves" mouthed by the surfer back at the beach. I remembered the signs warning about wolves at the gates to the beach parking lots. Have any of you women, knowing there might be a "pack of wolves" somewhere near, ever gone out in the middle of the night into the pitch blackness of a deserted forest . . . to bare your bottom for a pee? I was not too happy with the camping arrangement but, after the second venture out in the pouring rain, I became resigned. Yes, by then, it was raining again. Did I forget to mention that a good half-dozen centimetres of those three and a half metres of rain for which Tofino is renowned were now falling? Straight from the sky and quite steadily and heavily, I might add.

"Count your blessings, girl," my mother would have said. "You've got a good roof over your head." Yeah, sure, Mom, a van roof, that's all. I was in one big pickle.

First thing in the morning, Hans was up and outside, eagerly setting up the Coleman, coffee cups, and porridge pot on a makeshift table. In that squeezed-in little dry patch under a big cedar tree, as if by longstanding appointment.

"Are you planning to hike the seven kilometres back to the highway?" I gulped, cautiously blowing on the hot coffee to cool it down. Before he'd fallen asleep, he'd pointed the flashlight at the odometer.

"I set it when we turned this way," he'd said, as if I should believe his every word.

I suspected we were even further from the highway than he said. More likely, ten kilometres or more, and then we'd have to hitchhike another fifteen kilometres to Tofino or Ucluelet, whichever direction the first car along might be headed. Also, I was not having any luck dismissing the wolves from the overall picture.

"Not before we have a good hearty breakfast down the hatch," he laughed. No laughing matter, I felt like shouting, and then dismembering him, limb by limb. "You can stay here." The pack raced closer and closer to the lone, unprotected weak female, soaked to the skin and word-stricken. "Or come along with me," he added, as an afterthought.

"Doesn't look like we have much choice." I hardly had the words out when I heard a new noise. Oh, the welcome sound of a truck, a real truck rumbling over the Kennedy Lake Bridge. I dashed onto the road and in the direction from which the rumble came, arms windmilling for all they were worth. It was a one-woman protest, and no traffic would get through the blockade, at least not until she was assured of a way out. Up until that moment, I hadn't known about the fish hatchery maintained nearby by the Tla-o-qui-aht people, but I was more than a tad grateful that someone had the responsibility of attending the fish daily. The truck stopped and the driver heard my tale of woe. Hans ambled over.

"Haven't got booster cables, so I can't give you a boost, and I can't call you a tow truck," the driver said. That sinking feeling that I'd had in the middle of the night resurfaced. "But I'll call my wife. She'll get somebody here."

I heard the welcome click on the truck's two-way radio and the casual summary of our situation. I think he said his name was Bob. Nonetheless, by the time Bob hung up, I had a dreadful knot in my gut. Maybe he would drive away and we'd never hear from anybody again. Maybe we'd be no further ahead than we'd been when the owl flew off the alder branch. But he did not do that at all. Instead, he settled back in his truck, as if to wait for the end of the story. He was quite impressed that we'd seen a western screech owl up close.

"You ever see a sasquatch?" he asked. Oh, my God, I thought, first I had wolves to worry about and now I have the sasquatch. I didn't want upmanship; I wanted a battery boost.

"My granddad, he was the chief here about 1920. Twice, he saw a sasquatch. Once, he was over in the Fraser valley, but the second time, it was just east of here that the sasquatch ran across in front of him, maybe halfway to Alberni."

I decided not to listen.

"Other guys seen them, too."

One predator was enough and, besides, I wanted only to hear the buzz of the radio. It sounded finally, but the call was not from his wife; rather, from a truck driver on the way to haul out some heavy equipment. There was mumbled conversation for a few minutes and then I heard, "If it ain't off the road when I get there, I'll push it off." This road had more traffic than I'd originally thought.

To my relief, his wife buzzed almost as soon as that conversation ended. I kept one ear peeled for the truck bent on clearing the way for his outgoing load.

"The tow truck's on its way," Bob assured us. "I'll wait around till he gets here to make sure you people are on yours." I could have kissed the ground beneath him.

He started telling us about his kids and how he hoped they'd stay on the reserve. The radio interrupted us again, and then I heard the now-familiar sound of tires going over bridge.

What happened was this: a pickup showed up first, coming from town, and the driver had booster cables, so he boosted the battery and we began driving away, so as not to give the battery a chance to go dead again. We honked a thank you to both of them, the two good Samaritans in the Clayoquot forest just as the truck and flatbed rolled around a corner behind us. It passed us as soon as we'd made it over the bridge. We drove a few more kilometres down the road and met the tow truck.

"We can give you a cheque," Hans told him.

"Forget it. No trouble. I wasn't doing anything anyway." He pulled a U-turn, scattering a bit of gravel and sped back to Tofino, well ahead of us. The van purred along, as happy as a stroked kitten.

"Who was the guy who actually gave us the boost?" Hans asked.

"I'm sorry, Hans. The sign on the truck was half scratched off. I couldn't read it, and I didn't hear the man's name."

"I'll be," Hans sighed. "I hope somebody helps him out of a bind someday."

But never mind about all that. Now, when somebody says, "Clayoquot," I see a western screech owl. It is that straightforward and simple. That bird plucked my heartstrings like a newborn does its mother's. My Clayoquot owl.

Commuting by Kayak

– BONNY GLAMBECK –

"You're paddling home? You two are crazy!" Our friend shook her head as we geared up. She couldn't imagine kayaking after a long day of work, much less in the dark. This is still a common reaction from people in Tofino, although Dan and I have been commuting by kayak to our island home for fourteen years. It takes us a half hour, sometimes longer, to paddle to our cozy cabin nestled in the rainforest, but the trip is its own reward and always worth the effort.

On this night, I poked my head out the door of the waterfront office from which we teach kayaking. Shivering in the brisk evening air, I surveyed the conditions. The flag above the parking lot fluttered in the light breeze — stars would soon appear in the

Reprinted by permission of the author. Originally published in Sea Kayaker *magazine (www.seakayakermag.com).*

indigo afterglow of the sunset. Better dress warmly, I thought. I layered up, attached the waterproof flashlight to my personal floatation device, or PFD, and turned it on to check that it was working. It was already dusk as Dan and I loaded our groceries into the boats, but we'd be home before it was completely dark.

As we paddled, the light northeast wind began to gust, and blasts of cold air ruffled our hair. As it picked up, we speculated about this unusual wind, the direction and alarming rate at which it rose. Then we heard it hurling across the darkening water. In the distance, the sea's surface rippled black as a wall of wind came our way. The blast hit us suddenly, and I leaned into it, clawing forward. Wind waves broke over my bow sending cold spray into my face. Fear sent my mind racing through scenarios. What if the wind kept increasing? Within five minutes the conditions had changed from easy to the outer limits of my ability to move forward. Was this the front end of a squamish, a dangerous high-velocity katabatic wind? Reminding myself to breathe, and glad that Dan and I make a habit of paddling close together, I focused on my forward stroke.

We ferried across the current toward the beacon that sits off the shore in front of our cabin, but we were treadmilling — paddling as hard as we could but barely moving. As we strained forward against the wind, our boats parallel, the outgoing tide swept us off-course, downstream of the beacon. This gave us the chance to work our way up the eddy and peel out for the final glide home, our boats bouncing over the standing waves. In the seconds between the heavy gusts of wind, we sprinted forward into the darkness. Finally, in the lee of our island, muscles burning, we paddled up to the beach. Sliding our boats out of the water onto smooth gravel, we were soaked and rather stunned at how quickly a routine commute had turned into a frightening paddle.

I was first introduced to kayaking by Bruce, a friend who lived on a floathouse in the Broughton Archipelago off British Columbia's west coast. He used his kayak to visit the neighbours, pick up his mail at the general store and even salvage logs. For him, the kayak was first and foremost a vehicle to get around his water-based community.

Shortly thereafter, I found myself living on a small island near the village of Tofino. After searching for an affordable vehicle to transport supplies and myself to and from the island, I found that a kayak was the natural answer. I loved paddling and jumped at the chance to make it part of my lifestyle. Since my lack of experience and paddling partners, as well as the extreme nature of winter weather, often kept me stuck in town or trapped on the island, I also bought a motorboat. I still paddled much of the time, and as my skill improved, was able to handle more challenging weather. When Dan moved to the island in 1993, I gained a partner committed to paddling, and the motorboat fell into disuse, eventually to be sold.

I've paddled the same commuter kayak for the last fourteen years. It is a simple Kevlar model without gelcoat, bulkheads, hatches or rudder. I use air bags for flotation. My thirty-litre waterproof pack or a forty-pound bag of kitty litter slide in easily through the large cockpit. It's light enough when empty that I can flip it onto my shoulder to carry up the beach to its resting spot above the high-tide line. *Rockhopper,* a name the boat gained after many close encounters with rock gardens, looks like such a beater with its multitude of scars and patches that I don't worry about theft when I leave it "parked" in town.

Commuting by motorboat was certainly faster and easier. But faster isn't always better. Like the car on a daily commute to work, the motorboat moved me through the natural environment at a speed that disconnected me from the world around me. Now when I paddle home, the pace of the kayak allows me to slough off the buzz of my workday, slowing me down to what we call island time. When I head back to town, the kayak's tempo eases the shock of returning to the stress and demands of modern life, which even in our little village can be far too fast-paced.

The local vessel traffic also plays a part in this transition. I often feel like a rabbit crossing a busy highway, especially at night. My senses strain as I listen and look for fast-moving motorboats. Once I've cleared the traffic lane that runs along the waterfront of Tofino, my pulse slows, my breathing deepens, and my senses open to the wildness around me. Each morning as I head in to work, the town is a distant hum of activity. Paddling in, I'm eventually immersed in that hum. As I watch for boat traffic, greeting folks I know, my mind turns to the day ahead. Sometimes I'm eager to get on the water, and sometimes I'm not — but paddling every day keeps me in touch with the moon, the tides and changes in the weather. Information about the currents, tides and wind helps me choose which route I take so I can avoid dangers. Paying close attention to the seascape not only helps keep me safe, it keeps me connected to the spiritual element of wildness.

As a counterpoint to an increasingly troubled world, Clayoquot Sound holds out the possibility of healing and peace. As Renée Askins wrote in "Shades of Gray" (*Intimate Nature*, 1999), "We sleep better and dream deeper knowing there is a little wildness nuzzling at our door." In this sense, my commute is like a daily spiritual practice. The elements and animals don't allow me to stay lost in my own thoughts for long. Whether it's the adrenaline rush of paddling in a northwest gale or the silent dark eyes of a harbour seal, I'm always drawn back into the moment.

Paddling to town on my own one spring morning, I was essentially asleep at the wheel. Startled by a deep, resonant exhalation directly behind me, I glanced over my shoulder. A heart-shaped plume of mist hung above the water. The mottled tail of a grey whale emerged where my stern had been seconds before. The huge tail gently disappeared beneath the calm water leaving a slick "footprint" on the surface. Although the world had appeared mundane that day, something mysterious lurked beneath the surface. This is one of the most intriguing and sometimes frightening aspects of kayaking for me.

Another time as I paddled alone, a big tide was ebbing as I set out for town. To take advantage of the back eddy, I paddled a few feet from shore. Low water had exposed the colourful community of intertidal life. Iridescent seaweed shimmered in the bright morning sun; red tube worms and brilliant purple and orange starfish clung to the black rock. Out in the main current, I heard a loud snort and caught a glimpse of a Steller sea lion as it dived, riding the current toward me. I stopped paddling to watch this animal pass by. Sea lions can weigh two thousand pounds, so I wanted to keep an eye on this one.

It surfaced abeam of my boat, only fifteen feet away. Instead of coming up for air and carrying on, it stopped, turning its massive bear-like head toward me, mouth agape, exposing its large yellow canines. These sometimes aggressive animals are thought to carry their territory with them. We surveyed one another, until my proximity to this large carnivore unnerved me, and at that moment he dived in my direction. I sprinted for a nearby beach, powering forward, ready to brace. By the time I reached the shallows, the sea lion was nowhere to be seen. With my heart still pounding, I continued on to town. This was road rage of a sort that few commuters anywhere have to deal with.

When commuting by car, most people take little notice of the wind direction or speed, whether it's a full or new moon and whether the sky is clear or cloudy. Just getting from home to work can leave many people feeling over-scheduled and stressed. It's challenging to find time during the day to connect with nature, even for a walk or bike ride. And just like an extended paddling holiday in the wilderness, my commute leaves me refreshed and invigorated. It's a mini-holiday — a half-hour of uncluttered simplicity. But as with any holiday, there is preparation.

On one occasion, I was tired after Dan and I had finished teaching a six-day kayak course. After the trip cleanup and office business were taken care of, we prepared to paddle home. As we changed into our paddling clothes, geared up, packed the groceries into waterproof bags and carried our boats to the beach, it all seemed like too much effort. To top it off, it was dark and foggy. I envied the car commuter. As we left the beach, Dan tried to make conversation but I only mumbled monosyllabic answers.

The harbour lay shrouded in a thick layer of fog. As we left the shoreline, a curtain of fog closed behind us. Streams of bioluminescent organisms marked our bow wakes, tiny green stars streaming outward as our boats and paddles agitated the water — a common, yet magical occurrence on dark summer nights like this. Slowly we emerged from the fog. Above us, stars filled the black sky; behind us, the lights from Tofino lit the fog bank. We stopped paddling when we reached the shallow water over the mudflats. There was something different about the light from town. I realized that the radiance lighting the fog came not from the town but from the water itself. With the new moon, a big spring tide was powerfully flooding. The movement of the strong current was triggering the bioluminescence in the water.

I held my hand out over the water. My palm was illuminated by the weird glowstick-green light. We splashed the surface with our paddles, creating sparks of brighter green against the glowing background. Fish darted away in a lime flash. My workday blues were forgotten with the thrill of this magical phenomenon. We carried on, working our way up the eddy behind the beacon. As the water roiled and streamed around the beacon pillar, green light shone from the depths, like a swimming pool lit up at night. As I hit the top of the eddy, the current grabbed my bow, and I was flushed sideways downstream over a bull-kelp bed. Green sparks streamed off the giant kelp. It was dizzying to look down the stems trailing away into the depths, like flying through the Milky Way. Exhilarated — like children playing in the first snowfall of the year — we exchanged excited yips and hollers. We paddled back up to the top of the eddy, peeled out and rushed sideways over the kelp bed again and again. Commuting by kayak has its challenges, but these unexpected moments make it all worthwhile.

One morning, Dan and I carried the boats down to the water as we prepared to go to town. Walking back up the beach, I saw the grisly remains of a young raccoon. Something had made a meal of this little critter some time after the last high tide. Patrolling the area for a clue as to what had happened, we discovered wolf tracks in the sand parallel to shore. Following the tracks, we realized that the wolf must have been cruising the beach when it encountered the raccoon hauling its morning catch up from the sea. At the intersection of their tracks lay what was left of the wolf's breakfast. We lingered in the warm sun, basking in the thrill that a wolf had been on our beach only hours before. Tempting as it was to follow those wild tracks, the tide was rising under our kayaks; it was time to go to work. Traffic, fortunately, would not be heavy.

REFERENCE

Askins, R. (1999). "Shades of Gray" in B. Peterson, B. Peterson & D. Metzger (eds.), *Intimate Nature: The Bond between Women and Animals.* New York: Ballantine Books, pp. 375–378.

Bright Solstice Darkness

– FRANK HARPER –

In which I journey in deep-winter darkness toward home and get directed elsewhere by unknowable forces, learning that what is supposed to happen happens, and what's not supposed to happen happens too, because it's supposed to.

I was stuck in town, in Tofino, that week in December, many years ago. For days all my desires and even some of my energies were focused on getting away from the Christmas hype, on going home, to Island Beach for winter solstice. I was hoping to celebrate the return of the sun at my cozy, hand-built house, even if it meant I'd be celebrating alone. But that fickle, erratic friend, the weather, was not encouraging. Because I paddled a canoe in those days, I hoped for reasonably calm waters for three hours in order to complete the seven miles safely to my home-shore at the foot of Catface Mountain. But I wasn't getting those three hours: it blew southeast then southwest then southeast for a number of days and nights, heavy with gusts and chop and rain. On the day before the solstice it switched to westerly strong and steady, and my intuition — or maybe it was simply my desires — told me I'd be able to head for home that night. So I did my final shopping and packed my stuff down to the canoe at Jensen's Dock and loaded up in the late afternoon, in cold wind and sunshine.

I kept my weather-eye on the sky and water. Sure enough, the wind dropped off considerably, noticeably, not long after sunset, and by dark the harbour was flat and calm, the lights from Opitsaht blinking on the inky water.

I ate a Schoonerburger, then practically ran down to the dock, being pulled by some half-mad measure of homing-pigeon anxiety — now is the time! seize the moment! — and I untied the canoe before I realized that the paddle was under the boxes and bags of stuff, enough stuff to keep me at home through solstice, through Christ's birthday and right on through New Year's Eve. By rummaging and rearranging, I managed to pull the paddle out from under, and after I stroked into the outgoing tide I fished in my backpack for a flashlight and set it atop the gunnysack of clean laundry.

But I didn't need the flashlight right away. There were no other boats on the water that night — or none that were near me — and the sky was a sieve of winking stars. I headed right across the sandbar, feeling strong, digging the paddle deep, separating water. The air was clear and cold but I was so thoroughly bundled and was paddling so steadily that I was plenty warm. I remember letting out a series of jubilant yelps, releasing the pent-up energy that I'd been repressing in town for so long.

In Father Charles Channel, the tide against me, I dropped to my knees to paddle, and my hands were growing chilly right through the wool mittens, now wet with salt water. I could care less, I was going home.

Then a curtain of cloud unrolled itself across the starlight, shrouding me in a black cloak, a cape of darkness. I could, however, still make out the form of Lone Cone and — since I was right next to it — the shore of Vargas Island. Here I was, wanting to celebrate the Returning-of-the-Light, and all it did was get darker and darker. But it was quiet. I could hear only my breathing and the slurping of paddle-dipped water. I watched hundreds of

Frosty Esowista Peninsula. (PHOTO: JEN PUKONEN)

Frosty west coast walk, Flores Island, Clayoquot Sound. (PHOTO: JEN PUKONEN)

sparkling jewels flashing around the paddle as I stroked through a soup of bioluminescence.

Hugging the Vargas shore, about halfway to Catface, the canoe almost bumped into a low-hanging cedar tree that jutted out just above the water, so I interpreted that as a sign. I was (and am) a great follower of signs — signs from Somewhere Else, from the Unknown — because of their very mystery, because they require my own interpretations. And this sign told me to tie onto the tree and take a rest. Of course I obeyed.

I must've set the paddle on the seat of the canoe when I moved forward to find the line to tie to the tree. The line was tucked in, ship-shape fashion under the boat's bow-piece, and I unfolded my legs and made my way forward over the pile of cargo. As I did, my cramped knees buckled and my feet shifted in the small space and the canoe tipped at a precarious angle and, disoriented, I almost went overboard. But I grabbed the gunnels and regained my balance, and therefore the canoe's, and I pulled out the bowline. More carefully now, I tied the line to a cedar branch and made my way backwards to the seat where a banana and candy bar waited.

The paddle was gone.

The water was as black as the sky, as the shore. I could see nothing floating. I could barely see the water. I groped for the flashlight and aimed its meagre beam at the inkiness: dark liquid in motion. With a sense of rising panic — or accumulating adrenaline — I groped around the canoe. Maybe it was still aboard. Maybe it had squeezed itself back under the boxes. But no.

Still it was all right, because, being a safe boatsman, I'd brought along a spare paddle. So I again set about moving boxes and bags until the flashlight zoomed in on the spare which I tugged at and set carefully between my legs as I chewed the candy.

This paddle was a spare. It had drifted into Island Beach after a storm that winter, and I'd found it half-buried in the sand, obviously a gift from the sea, obviously meant for my use. I remember thinking, "Fate has its ways," as I dug that carved paddle out from the sand. And it does have its ways. Now I untied the bowline and draped it across the cargo for easy access, and against the current, I plunged this paddle into the water, deep. The boat responded at once. I dug that paddle two more times and then I heard and felt the shaft crack right in my hands. *Oh, no no no! No!* I could feel splinters through my mittens. It was broken in the worst spot, at the base of the shaft. I turned the paddle around and tried to use it gingerly. The canoe moved maybe a foot before the rest of the paddle broke.

I managed to grab onto the ghostly cedar and I tied on again. I don't know how long I sat there, tied to that tree in the darkness. I know I quickly grew cold. I moved around as much as I could every so often to try to warm up. I know I ate more bananas. I know I found the whisky and had a sip. Then later, another sip. But it made most sense to do nothing. Getting onto shore wouldn't help much, and I could do that later if I wanted to stretch, or climb along the rocks . . . to where?

I heard a humming, the droning of a fishboat's engine as it rounded the far point of Vargas, and at first I thought it was heading across to Lone Cone. But no. I listened hard. It was humming along in my direction. I was sure of it. The humming was coming closer. Finally I could see a churning wake, bubbling, spreading, and I could see the boat's silhouette. I shone the dwindling flashlight in the boat's direction: the beam still cast a faint glow. I shook the flashlight up and down, around and around, waving it. The fishboat came closer but it wasn't slowing down. They *must've* seen me.

I yelled, "Hey! Hey there! Help!"

The boat cruised steadily on its course about thirty yards away, ploughing toward Tofino, the diesel engine rumbling right along.

I did what you're not supposed to do. I stood up in the canoe splay-legged and waved my arms and waved the flashlight and yelled more, and the fishboat's engine suddenly cut out, dropped its RPMs, purred, and the boat slowed.

I heard a voice call out, "Is somebody there?"

"Yes," I yelled. "It's me!"

"Who's me?"
"Me. Frank."
"You okay?"
"No. I've got no paddle."
"Neither do I," called the voice. "You need a tow?"
"I guess so. Yes. Yes. I do."

So the fishboat edged in close, slowly, and I used the paddle's blade to scoop alongside. As I climbed aboard the fishboat, a man in the dark helped me tie the canoe onto a rail, and he laughed, "It's dark out here, too dark for you to be resting." I then recognized him as Peter Charlie from Ahousaht, now deceased.

"I'm going to Tofino," he said.

"Okay." I accepted my fate.

In the boat's cabin, rocking on a rickety stool, Peter Charlie reminded me, "The last time you were on my boat was summertime."

I agreed, remembering. It had been years earlier, right around the summer solstice in fact. Peter Charlie had given me a ride from Tofino to Catface on my first trip to Clayoquot Sound, the journey that introduced me to the place, to my new home, another journey that changed my life.

"Maybe we're hooked together," laughed Peter Charlie as he gave the fishboat more speed.

Defending the Forest

Chronicles of Protest at Clayoquot Sound

– JOANNA L. ROBINSON & DAVID B. TINDALL –

The last stands are quietly logged,
leaving no memory. We cannot see
anymore. We all have no idea of
what beauty we have disappeared.

— Annabelle, Friends of
Clayoquot Sound member

Annabelle, an artist, mother and self-described "focused member of the planet," is concerned about the disappearing forests on Vancouver Island. She worries that we are no longer living in harmony with our natural surroundings, and thus cannot recognize and appreciate the immense beauty of nature.[1] As a forty-four-year-old mother, she wants to see the old growth rainforest of Clayoquot Sound preserved for future generations. Having grown up on Hornby Island, Annabelle feels strongly about practising a way of life that values the natural environment and is raising her child to love and respect nature. In response to our survey, she writes:

> We have to be in harmony within ourselves in order to be able to live with gentleness to the rest of the world. . . . I am bringing up my child, aged three, to love the wilderness.

Annabelle believes that through education and dialogue we can achieve social change. As an artist, Annabelle is devoted to painting wilderness scenes in order to show others the beauty of nature. She strongly desires a new way of living with the forest, where the trees are "preserved for their beauty and greatness." But Annabelle also believes that we must engage in collective action to defend the old growth rainforests from being clear-cut, particularly those unique and irreplaceable "last stands of old growth" on the west coast of Vancouver Island.

In the summer of 1993, Annabelle stood, along with hundreds of other protesters — parents, grandparents, children, loggers, business professionals, students, celebrities, rock stars and tourists — on the Kennedy River Bridge in Clayoquot Sound, as part of the largest episode of non-violent civil disobedience in Canadian history. She and the other activists, over 900 of whom would eventually be arrested, were trying to prevent the logging of one of the world's largest remaining temperate rainforests. Annabelle describes this experience as part of her commitment to "practise living what one wishes others to see; how to live upon this planet with reverence for the environment and community."

Annabelle's story is not unlike thousands of other protesters — over 12,000 in total — who travelled to Clayoquot that year in order to stand in defence of the old growth. It is a story of activism. It is a story about perseverance in the face of powerful foes, a story of individuals from the far corners of the planet who challenged the dominant modern narrative of nature as a source of economic wealth, and demanded a new way of living in harmony with the natural world. During those months of protest in 1993, Clayoquot Sound — with its rare pristine (completely unlogged) watersheds, mountains and lush valleys full of giant moss-covered trees, some believed to be as old as 2,000 years — was, for the thousands of protesters who made the journey, a place of sanctuary and peace, a place of spirituality, a place called home.

A HISTORY OF PROTEST

British Columbia, with its vast areas of untouched wilderness and its resource-dependent economy has, not surprisingly, experienced a fair share of environmental conflict. The province is home to one quarter of the world's remaining ancient temperate rainforests. These forests have often been the site of battles waged between environmental groups, aboriginal peoples, forest industry workers and wilderness lovers, among others. The public became progressively more concerned during the 1980s and 1990s, as environmental organizations increasingly argued for the preservation of the remaining temperate rainforests on Vancouver Island. A campaign of civil disobedience began with a series of direct-action protests aimed at preserving old growth forests, including the battle to save the Carmanah and Walbran valleys and the logging blockades on Meares Island and at Sulphur Pass. These direct action protests helped raise public awareness and international media attention over clear-cutting practices by the forest industry in B.C.[2]

The climax of the forest preservation movement here occurred in the summer of '93, with the protest over logging in Clayoquot. That April, the provincial government announced plans to open up two-thirds of the Sound to extensive logging, sparking the peaceful blockades that garnered international attention from the media and environmental organizations. Soon there were protests around the globe; B.C.'s rainforests became a global symbol for conservation.[3]

COMMUNITY CONCERNS ABOUT FORESTRY AND CONSERVATION ON VANCOUVER ISLAND

Why did thousands of people travel to Clayoquot to stand together on a logging road? What is it about the unique forest in British Columbia that changes the way people view nature and how they mobilize for change around environmental problems? These are just some of the questions we wanted to explore as part of our research on participation in B.C.'s environmental movement.

Because the Clayoquot protest was a pivotal event in the history of the environmental movement in Canada, we were interested in looking more closely at this episode of collective action, in order to better understand why people participate in the forest preservation movement. The research and stories we share are from a survey that is part of a larger project examining people's participation in the environmental movement in B.C. and Canada, focusing on the members of the Friends of Clayoquot Sound (FOCS), one of the key players in the protest.

Located in Tofino, FOCS was the organization that launched the blockade at Kennedy Lake Road, and coordinated a major international campaign to protect the ancient forests of the region. Its approximately 500 members come from different regions across the province, the country and the world. Established as a non-profit society in 1979, FOCS is a grassroots, community-based environmental group with two main goals, one biological and the other cultural. One goal is to preserve natural biological diversity and ecosystem dynamics, concentrating on Clayoquot Sound's ocean and forest. The other goal is to stimulate creation of a conservation-based society, with a corresponding conservation-based economy. "Our organization is part of a global conservation movement calling for a shift away from unsustainable practices that damage the health of earth's ecosystems," states Maryjka Mychajlowycz on FOCS' website.

As part of our study, the authors of this paper mailed a self-administered questionnaire to the members of the organization. In total, 212 completed questionnaires were returned in 2005, a response rate of over 50 percent. Although the survey was composed of mainly closed-ended questions, there were several open-ended questions where people filling out the survey were invited to express their concerns about forestry and conservation on Vancouver Island. These written responses form the basis of these narratives of home as they shed light on the relationship people have to the rainforests of Vancouver Island, why they want them preserved, their concerns about forestry practices, and proposed solutions to the problem of deforestation in Clayoquot Sound and elsewhere in British Columbia.

METAPHORS OF THE FOREST

We understand nature through significant social and cultural symbols that assign meaning to the world around us and inform our relationship to the environment. These symbols or metaphors shape our values, beliefs and understanding of nature and the way in which it should be treated.[4] Many FOCS members use metaphors to describe their attachment to the rainforests, evoking a sense of spirituality, peace and sacredness to justify the preservation of the forest for its own sake. Although many of them live hundreds of miles away, their descriptions evoke a strong sense of belonging and a deep emotional attachment to this place.

Janet, a forty-five-year-old political activist who lives in East Vancouver, has spent years mobilizing to protect these forests, places she considers sacred. She believes in the power of political protest to bring about social change and has organized and attended teach-ins and open discussions about wilderness conservation. Like Annabelle, she tries to educate others about being environmentally responsible through her lifestyle choices; she recycles at home and at work, buys local, organic produce, regularly helps maintain parks and natural habitats and doesn't own a car. In the past, she also wrote letters to international companies as part of FOCS' Markets Campaign — an international initiative to pressure logging companies to stop the clear-cutting of old-growth by educating consumers and the major corporate buyers of forest products. Janet believes it is criminal to log in Clayoquot, as it is a sacred place. A spiritual and emotional attachment is her motivation for action.

Anne, a forty-three-year-old long-time member of FOCS, strongly identifies herself as a member of the wilderness preservation movement; she regularly engages in discussions about conservation issues and belongs to several prominent B.C. organizations. She was among the blockaders. Although she now lives and works in Vancouver, Anne grew up in the Slocan Valley, an area known for its magnificent mountain ranges and glacier river valleys. Protecting nature is a value she holds dear; in her opinion, when we disrespect the natural world, we disrespect ourselves. She explains:

> Treating forests sustainably is also important to humans for "spiritual" reasons. Our self-respect, sense of wonder and connectedness to the whole universe benefit when we act with appreciation, and suffer when we are destructive.

Anne believes in the importance of balancing the global ecosystem and preserving forests' biodiversity. She considers the current economic model a threat not only to nature but also to the human spirit.

Others who answered the survey see Clayoquot as a unique place, whose ecological integrity is threatened by clear-cut logging practices. Wendy, a sixty-three-year-old botanical illustrator, also lives in Vancouver. When asked if she had any concerns about forestry and conservation on the island, she answered:

> Absolutely! I believe the forest on the west coast of Vancouver Island is a sacred world treasure and the many layers of wonderful vegetation are completely destroyed by clear-cut logging. . . . The coastal ecosystem . . . is unique and irreplaceable.

Wendy is adamant that ancient forests be preserved for posterity because of the unique resources they provide as standing forests. Part of what makes Clayoquot Sound special and unique to Wendy is biodiversity. She explains: "As a botanist, I am worried about loss of biodiversity — another resource not enough appreciated. I think we are selling our birthright off too fast and too cheaply!"

For Wendy, clear-cutting is a serious threat not only to the biodiversity of the region, but also to its ecological integrity, which demands the maintaining of stream temperatures for fish spawning and the protection of pristine watersheds. As an avid mountaineer and sea kayaker, Wendy sees the forest not just as a habitat for wildlife, but also as a place of recreation:

> Also it is the prerequisite for recreation enjoyment! . . . I am very much interested in maintaining as much old growth as possible — being an avid hiker I find dog-hair second-growth incredibly depressing and think the government should preserve as much old growth for recreation and its own sake as possible — an irreplaceable resource for the future.

In a province where much of the forests has been logged, replanted, and logged again, there is something unique and precious about old growth, with its many layers of vegetation that veil the forest floor. Wendy sees a greater value in preserving the old growth than logging it for profit. The services it provides in terms of habitat, maintaining ecological equilibrium and a place for recreational enjoyment trump the traditional use-value of clear-cut logging.

Many FOCS members who answered the survey, particularly those who participated in the 1993 protest, express a sense of ethics when describing the way they view nature, ascribing a sense of our moral responsibility to the forest that justifies its protection. Barbara is a seventy-six-year-old retired professor who was arrested at Clayoquot in July of '93. She was charged with trespassing and sentenced to twenty-eight days' house arrest. During the trial, she read a statement to the court, a cautionary tale of democracy and responsibility. A copy of this statement was attached to her survey response. In it she describes the moral imperative of individuals to protect that which sustains life on the planet, claiming that it is the responsibility of citizens to defend the forest from clear-cut logging. She argues that in the face of danger — in this case the systematic destruction of the last old growth forests in Clayoquot Sound — the protesters were simply acting in response to an ethical duty that requires humans to protect that which provides them sustenance.

Carole, an eighty-four-year-old retired legal secretary from the Comox Valley, also participated in the blockade, as well as many other protests around B.C. Like Barbara, she sees those who engage in civil disobedience to prevent clear-cutting as practising a moral duty, while those who clear-cut are only interested in profits: "destroying their own livelihood by cutting themselves out of wood." Carole believes that decades of destructive logging practices — including pesticide use in watersheds and lack of stream bank protection — have devastated forests here. As a result, more and more people engage in and support direct action to stop these practices:

> I can say from experience that no one would be blockading logging roads . . . nor boycotting products, if we had sane logging practices in B.C. I have been on blockades on several occasions — Tsktika, Walbran, Clayoquot Sound, and sundry smaller ones over logging company rape of private and publicly owned land, and I helped prevent further logging and mining in Strathcona Park in 1988. We style it CIVIL OBEDIENCE, i.e. when the powers that be fail to protect our environment, then those of us who care have not only a right but a duty to move in and save all we can.

Carole adds a new dimension to the reasons for engaging in direct action: the failure of government policy to look out for the long-term interests of the environment and resource-dependent communities of the province. She firmly believes that we must demand better conservation policies from our elected representatives and has lobbied all levels of government for over twenty years, writing letters to Members of Parliament, signing petitions and demonstrating in front of the Parliament Buildings in Victoria. Those who participated in the blockade, as defenders of the forests, consider such protection a form of democratic responsibility and an exercise in citizenship.

SUSTAINABILITY NOW

Although most FOCS members value the temperate rainforests for something other than economic profit, as demonstrated in the above narratives, many also believe in maintaining the economic use of the forests, albeit in a sustainable way. Those who value the economic importance of the forest are critical of current logging practices, and the economic model that encourages clear-cuts and rapid deforestation to satisfy an increasing demand for profits. Jim, a fifty-two-year-old teacher living in Vancouver and long-time FOCS member, believes that most environmental problems are caused by overconsumption and the unsustainable use of natural resources. According to him, British Columbia's vast wealth of resources belongs to the people who live here, and not to the shareholders of "a company that has been granted a license." He advocates a new economic model that values what is best for balancing the needs of the ecosystem with the needs of the communities whose livelihood depends on the forest:

> I believe we must move to an economy based on sustainability and meeting needs. . . . Our economies need to be focused on meeting physical needs (food, air, water). . . . Local production and consumption of products grown/made here need to be emphasized to reduce huge amounts of pollution generated due to transport which is taxing our forests' ability to cleanse.

A more general critique of macro-economic structures and a globalized consumer capitalist economy is what drives Jim's activism. The potential clear-cut logging of the area has become a symbol of B.C.'s complicity in this economic system, and the race to fell the last tree for profit. According to Jim, there should be no more logging in old growth areas or watersheds. He also argues that all logs from second-growth forests in the province must be processed *locally* into finished wood products. These products would generate higher revenue for forest-dependent communities; investment in value-added industry would greatly reduce the number of trees cut as well as create long-term employment in the communities that have been hard hit by mill closures.

Many others also argue for a different kind of economy in B.C. For example, Sue, a fifty-one-year-old FOCS member who lives in North Saanich and works as a freelance natural history writer, echoes Jim's call for a more sustainable way of life. Sue has lived on Vancouver Island all her life and grew up in a forestry-dependent community. She describes the immense economic and environmental changes that are the result of what she terms "poor forestry practices" — clear-cutting, soil erosion, slides, extreme damage to streams:

> I have witnessed the enormous changes to the local economy, some of them positive and many of them negatively impacting the natural environment. It has been heartbreaking to see the damage wrought by the force that drives our North American way of life — corporate profit — coupled with the belief that we must have constant "growth." Why not sustainability?

Tree-sitting protestors and their supporters were successful in saving the 800-year-old Eik St. cedar tree in Tofino from being cut down, April 2001.
(PHOTO: WARREN RUDD)

Strawberry Island, Clayoquot Sound. (PHOTO: JEN PUKONEN)

A greater tragedy for Sue is the damage being done to the most remote and pristine forests on the west coast of the island, in Clayoquot and beyond. She believes that these areas are particularly vulnerable to clear-cuts because they are so remote and thus less easily monitored by government and environmentalists. She argues that a new way of practising forestry, based on sustainability and value-added industry, would go a long way in preventing further destruction.

Others pointed to the short-sightedness of focusing on short-term profits over long-term sustainability. Janet, an eighty-three-year-old widow and retired nurse, believes that an economic system whose priority is "financial gain for now" is both an abuse of nature and a threat to the preservation of the forests for present and future generations. As a long-time island resident who believes in supporting the local economy, she is dismayed with an economic system that supports "destructive methods, such as clear-cutting . . . to satisfy the financial gains of a few." Because of her concern for future generations, Janet travelled to Clayoquot in '93 to join the blockade, where she was arrested.

> If we abuse nature . . . we and our successors will suffer. We are a part of nature — city dwellers sometimes forget this. We need the trees, globally, to turn carbon dioxide into oxygen. Canada has signed the Kyoto accord.

Janet explains that selective logging methods have already been practised on private property, such as Merve Wilkinson's Wildwood, near Nanaimo, and "have proved that forests can remain and produce indefinitely."

As the current model of economic development in B.C. threatens places like Clayoquot, and other remote, pristine forests, people have been mobilized to think critically about the model and to dream of less destructive alternatives. David, a physician, researcher and father of four from Vancouver, is seriously concerned about current forestry practices. Although he believes in the importance of a strong economy and the vital services it provides, such as health care and education, he does not think that the natural environment should be a source of economic wealth and jobs. Rather, he believes that the focus of the new economy should be on the high-tech industry. Instead of being logged, forests should be preserved so their natural beauty attracts people to the province for tourism or employment:

> I believe . . . all areas of potential tourist value should be spared. I am not sympathetic to preserving forestry industry jobs. Rather, we should be going for high-tech jobs in B.C. and attracting [workers] because of our natural beauty and preserved environment.

David proposes a new model for the economy — preserving nature to attract tourists and highly skilled workers — abandoning the old unsustainable model of short-term profit. For him, this means adopting a new value system in Canada, one that recognizes the importance of economic growth, while also preserving the natural environment for the health of the planet. He sees the reorganization of the economy, through investment in non-polluting, non-resource based sectors, as the first and necessary step to "developing a reasonable and sustainable balance between humanity and all other life on the planet." David joins others who see the ancient temperate rainforests as an irreplaceable resource, whose preservation could have long-term economic benefits for the province.

HOW DO WE GET THERE? ECONOMIC, ENVIRONMENTAL AND SOCIAL SOLUTIONS

Most people who answered the survey described their concerns about forestry and conservation on Vancouver Island. At the same time, they also wrote about their desire for a new way of life, one that recognizes the inherent value in the beauty and sacredness of nature. Many people offered what they see as reasonable and sound solutions to the problem of deforestation in B.C.

The most important step, according to most of the individuals who responded to the survey, is immediately and permanently to put an end to logging the island's old growth, three-quarters of which is already gone. Ed, a sixty-four-year-old retired teacher from Vancouver and long-time FOCS member, expresses his concern with dependence on such logging and calls for a new conservation-based policy:

> Far too large an area of the island has been logged off. The forest industry's focus on old growth timber threatens the few old growth areas left. The impact on the natural life of the forest by the logging industry has been immensely negative, especially on salmon-bearing water courses. There should be no further logging of any old growth forest. . . . The program of deactivating forest roads should continue.

Ed describes the key role of the old growth forest in maintaining the ecosystem of the region, and argues that clear-cut logging should be banned in order to protect precious habitat.

Many others echo Ed's call. Helen, a seventy-year-old widow and retired teacher, believes in setting an example of "joyful simplicity" in the way that she lives her life. As a long time resident of Nanoose Bay, she wants to preserve island forests for future generations. She believes in the power of education in bringing about social change. As part of this commitment to education, she regularly takes her grandchildren to the west coast of the island so "they can understand what they are losing." Although she still considers that "big business has the upper hand in the forest," she has hope that through legal channels, such as legislation to ban the export of raw logs, logging of ancient timber can cease:

> All land seems to be for sale, even Crown land. We do not need to cut any old growth forest! We should survive on second-growth. Export of raw logs should be illegal. Forest tenures should be allowed to lapse. I know the legal implications but "due notice" should be reasonable and then no more cutting on old growth. Government made laws and they can and should unmake them.

She points to the government's power to stop and reverse their previous forest policies in order to protect the old growth forests while allowing for logging in second growth forests.

On the other hand, many people report having less faith in government to regulate logging. Mark, a thirty-nine-year-old full-time father who lives on Bowen Island, is angered by the provincial government's complicity with the multinational forestry companies, both in terms of allowing clear-cutting and the disappearance of community-based forestry. He states:

> Our own Ministry of Forests has admitted that big business [is] cutting above the sustainable rate. We need to stop allowing our government allowing unsustainable clear-cut logging of our forests. We need to stop shipping jobs out of the province in the form of raw logs and timber, and utilize value-added markets while keeping the jobs here in B.C.

He claims that campaign donations from forestry companies prevent individual citizens from having a say in protecting the environment. Mark believes that investing in community-based small-scale lumber mills and value-added industry is the best way of ensuring the environmental and social sustainability of Vancouver Island communities. Rather than trust in government to lead us down the right path, he believes that consumer boycotts and educational campaigns about the dangers of clear-cutting are the most effective way to bring about this change.

BUILDING A FOREST COMMUNITY

The idea of community emerged as an important value to most FOCS members in their discussions of forestry and conservation on Vancouver Island. Many people expressed that a strong sense of community was one of the reasons they participate in the environmental movement in general, and in FOCS campaigns in particular. Furthermore, for many people, preserving small forestry-based communities is considered as important a value as preserving the forest — most people see these two values as inseparable. As such, the solution for protection and economic sustainability is an investment in community-based forestry practices. In their survey answers, people state that a switch to community-based forestry, with small-scale lumber mills, selective logging practices and value-added industry would allow for sustainable management of forests, as well as maintaining the social fabric that is needed for communities to thrive. While few believe that putting forests back into the hands of local communities will be easy, they believe it is the only way to put an end to clear-cutting and ensure that something is left for future generations. Perhaps Annabelle, the forty-four-year-old artist and "focused member of the planet," puts it best when describing the kind of communities British Columbia needs to maintain harmony and balance with nature:

> I think we need to have small communities interspersed within the natural place, filled with people caring for the surrounding environs, for example streams cleaned up for salmon, trees preserved for their beauty and greatness. Forestry working within, gently beautifying yet "harvesting" the needed timber for furniture. The wood stays close to home. Then, only then, is the product shipped off to the rest of the world. There would be a butcher, a baker, a hydro maker (candlestick maker). We would manage to make a complete community not a ghost town. It is just very one-sided to have just one reason such as trees — for world export — to keep a community.

FOCS members challenge the dominant economic narrative in British Columbia. People express a sense of respect, awe and reverence for the ancient temperate rainforests of Vancouver Island, particularly for the pristine forests of Clayoquot Sound. They see a forest not as a mere profit-generating resource, but as a different source of wealth — a place of beauty and spirituality, a place to experience joyful simplicity, a space for recreational enjoyment and a sustainer of small forestry-based communities. Their belief in environmental and economic justice has led them to adopt a different paradigm from the dominant one that views nature only for economic gain. Their love of the forest has moved them to protest, to educate, to write and to make what to them are pilgrimages to Clayoquot Sound.

Despite the overwhelming sense of anger, shame, fear, frustration and concern that many Friends of Clayoquot Sound members express in their narratives, the many proposed solutions and paths to a new life-sustaining economy offer some hope. Although these individuals have yet to see their visions of nature and the economy take centre stage in this province, what is clear from their stories is that Clayoquot Sound is "home" to thousands of people around the world, whose determination in the face of disenchantment will ensure that the struggle to preserve this rare ecosystem will continue for years to come.

EDITORS' NOTE: *Although Clayoquot Sound became a UNESCO Biosphere Reserve in 2000, logging of its old growth forests continues. The sizes of clear-cuts vary but can still stretch considerably, with odd clumps of trees left here and there in what the industry has termed variable retention silviculture. These clumps are particularly vulnerable to the blowdown effect and will become more so now that winds are increasing in this era of climate change. There is immense survival-value of ancient forests as carbon sinks to mitigate global warming. All forests offer this but none more than ancient stands. The largest area of ancient forest left on Vancouver Island is Clayoquot Sound.*

REFERENCES

Berman, T., G. Ingram, M. Gibbons, R. Hatch, L. Maingon, & C. Hatch. (1994). *Clayoquot and Dissent.* Vancouver: Ronsdale Press Limited.

Blake, D., N. Guppy, & P. Urmetzer. (1997). "Canadian Public Opinion and Environmental Action: Evidence from British Columbia." *Canadian Journal of Political Science.* 30(3): 452–471.

Breen-Needham, H., S. Duncan, D. Ferens, P. Reeve & S. Yates. (1994). *Witness to Wilderness: The Clayoquot Sound Anthology.* Vancouver: Arsenal Pulp Press.

Greider, T. & L. Garkovich. (1994). "Landscapes: The Social Construction of Nature and the Environment." *Rural Sociology.* 59 (1): 1–24.

MacIsaac, R. & A. Champagne. (1994). *Clayoquot Mass Trials: Defending the Rainforest.* Gabriola Island, B.C.: New Society Publishers.

Magnusson, W. & K. Shaw, (eds.). (2002). *A Political Space: Reading the Global through Clayoquot Sound.* Montreal: McGill-Queen's University Press.

McLaren, J. (1994). *Spirits Rising: The Story of the Clayoquot Peace Camp.* Gabriola, B.C.: Pacific Edge Publishing.

McNaughten, P. & J. Urry. (1998). *Contested Natures.* Thousand Oaks, CA: Sage Publications.

Robinson, Joanna L., D.B. Tindall, E. Seldat & G. Pechlaner. 2007. "Support for First Nations' Land Claims Amongst Members of the Wilderness Preservation Movement: The Potential for an Environmental Justice Movement in British Columbia." *Local Environment.* 12(6): 579–598.

Tindall, D.B. (2002). "Social Networks, Identification, and Participation in an Environmental Movement: Low-Medium Cost Activism within the British Columbia Wilderness Preservation Movement." *Canadian Review of Sociology and Anthropology.* 39(4): 413–452.

Tindall, D.B. & N. Begoray. (1993). "Old Growth Defenders: The Battle for the Carmanah Valley." In S. Lerner (ed.) *Environmental Stewardship: Studies in Active Earth Keeping.* Waterloo, Ontario: University of Waterloo Geography Series: 296–322.

Wilson, J. (1998). *Talk and Log: Wilderness Politics in British Columbia.* Vancouver: University of British Columbia Press.

NOTES

[1] In order to protect the identity of the respondents, all names used are pseudonyms.

[2] For more information on the wilderness movement in British Columbia, see Blake, Guppy, and Urmetzer 1997, Tindall 2002, Tindall & Begoray 1993, and Wilson 1998, Robinson et al. 2007.

[3] For more information on the Clayoquot Sound protest see Berman, et al. 1994, Breen-Needham et al. 1994, MacIsaac & Champagne 1994, McLaren 1994, and Magnusson & Shaw 2002.

[4] For more information on the social construction of nature/metaphors of nature see Greider & Garkovich 1994 and McNaughten & Urry 1998.

Contact Luna

- HANNE LORE -

When first I arrived at Yuquot, Nootka Sound, a decade ago, I sensed a powerful past as I walked up the long boardwalk to the Big House. I climbed a hillock near the church and was rewarded with a vista to the east toward Conuma, Sacred Mountain, where the bones of our ancestors were once taken. This is the place where over two hundred years ago Chief Maquinna, a great warrior and whaling chief, was laid to rest. It is said the chief's funerary vessel had been hollowed from a single giant cedar tree in the image of Kakawin: Orca, Protector of the Seas. Maquinna's casket was long ago taken by the elements, but his spirit is still strong in the land and in his descendants.

On my visit I met Hazel while walking along the shore. She was a sturdy woman in her fifties, had sleek black hair and dark, angry eyes. Our afternoon walk became a daily habit, and sometimes her granddaughter, Della, just four years old, accompanied us. Hazel would talk softly, and Della would answer, as I listened pensively to the sounds of a language I did not understand. On our walks, we ventured to many places near their home, including Jewitt Lake. Before contact, before "Jewitt," the lake had had another name, the calm dark waters the site of a whaler's shrine, on an islet about a hundred metres from shore. Hazel didn't tell me much about the rituals involved with the shrine, just that it had been a whaler's sanctuary, and that it once had life-size carvings of people — cedar images of those who had been sacrificed in pre-hunt whaling rites. The statues had been sold to an outsider by a local during the mid-1800s and were now in a museum in New York. Hazel told me that the Moachaht were negotiating to recover them, bring them home, maybe even return them to the shrine.

"How big were the statues?" I asked.

She didn't answer for a long while, her eyes fixed on the islet.

Then she said, "Don't swim in that lake."

One afternoon toward the end of my visit we sat on the western bluff amidst Indian paintbrush flowers, looking out over the open Pacific. Della was picking flowers while Hazel shaded her eyes against the low afternoon sun. I followed Hazel's gaze and noticed a pod of orcas gliding past in the distance, heading north, dorsal fins high.

"What are you thinking?" I asked.

"I'm thinking how special those creatures are. For us they are protectors of the ocean, you know, Spirit Whales. You used to see more of them."

"But you come from whalers' ancestry. Didn't your forefathers hunt those whales?"

She smiled patiently. "We never took the Kakawin, the orca. Kakawin were always sacred."

I returned to Yuquot in 2004 at a time when Muchalaht Inlet had been in the news for several months. The catalyst was a solitary young orca that had taken up residence in the sound. The cetacean was said to be male, perhaps four years old, identified by marine scientists as L98. He was subsequently named "Luna" by a romantic public-at-large. Luna was believed to be a youngster from the Southern Resident L-Pod, which are salmon-eaters rather than seal-eaters.

The whale liked to spend time around the Gold River public dock beside the abandoned site of the original Moachaht reserve. He seemed interested in communication with people, popping up to have his head stroked by human hands. He would sing, too. Marine scientists and public servants believed that his behaviour indicated boredom and loneliness — and that he should be rehabilitated to his original family.

A lot of money was raised on both sides of the forty-ninth parallel, and a holding pen was constructed in Muchalaht Inlet, stocked with sockeye salmon. Luna was to be trapped in the pen, examined by marine medics, then trucked to the east coast of Vancouver Island. He was to be kept in another pen somewhere north of Campbell River, and when L-Pod swam past, hopefully calling to the errant family member, Luna would rejoin the world of whales. That was the plan. Meanwhile, more and more local people were complaining that the creature was becoming a pest — that he was interfering with their economic livelihoods by damaging the propellers of docked fishboats and floatplanes. One of his specialties was the dismantling of sonar equipment.

On the day of the proposed capture, pandemonium broke out on Gold River dock, adding more excitement to the media circus taking place on the shoreline. Initially, L98 was successfully herded into the pen by Fisheries boats. I heard he sneaked in quickly, snatched a salmon, and changed his mind. Before they could close the hatch he was distracted by a rhythmic drumming from the dock. The First Nations people were out in full strength: there must have been sixty paddlers, women and kids too. Scores of paddles pounded a rhythm on the dock while a hundred human voices sang a soft, melodious song that sounded like prayer.

L98 took a sharp turn back toward the pier. The paddlers jumped down into two large canoes, ready and waiting. They led Luna between them, speeding off west into the dusk, and in the stunned silence that followed, the powerboats did not take pursuit.

Tsuxiit/Luna leaping from the water while members of the Mowachaht/Muchlaht First Nation lead the orca away from the Gold River dock by war canoe.

(PHOTO: GERRY KAHRMANN, *PROVINCE* STAFF PHOTO)

Members of the Mowachaht/Muchlaht First Nation lead Tsuxiit/Luna away from the capture pen towards the open sea.
(PHOTO: GERRY KAHRMANN, *PROVINCE* STAFF PHOTO)

Later that evening the paddlers returned without the whale. I overheard Mike Maquinna, hereditary Chief of the Moachaht/Muchalaht Nation, being interviewed on the wharf in the half-dark. He spoke of why the whale was important. Three years earlier, his people already knew the young orca and named him Tsuxiit — three days after Mike's father, Chief Ambrose Maquinna, passed away. On his deathbed his father proclaimed he would come back in the form of Kakawin. For many Moachaht, Chief Maquinna's spirit now resided in Tsuxiit. They respected the whale's choice to be here. They believed Tsuxiit knew what he was doing.

On June 20, 2005, I returned to Yuquot for a third visit, arriving at Gold River dock on the evening of the longest day of the year. It had been hot on the road, and I was weary. The media had decamped weeks ago. There was no movement or sound other than the stroke of wind passing over Muchalaht Inlet.

I wandered along the pier to the observation tent set up by the Nation. A woman offered me water and a cup of tea. She informed me that every day since the attempted abduction by Fisheries, canoes had been leading Tsuxiit back out into the Sound to keep him away from the public. She smiled and added, "The exercise has been good for the guys."

Tsuxiit had brought her community together. Because of him, people were remembering the past and taking a stand for value systems forged by their ancestors. The legacy of Chief Ambrose was taking effect; Tsuxiit was the deceased chief's way to unite the people.

Early the next morning I boarded the *U'chuck* for the journey to "Friendly Cove," so named by James Cook. Many Moachaht and Muchalaht families were on the ship, on their way out west. All eyes searched for a glimpse of Luna, but he was nowhere to be seen.

I stepped off the boat with the same sense of magic I remembered. Near the Big House, the Yuquot Totem, already leaning a decade ago, had tumbled onto shore grass in a cycle that honours nature's way.

Hazel had aged gracefully, her once black hair now sprinkled with white. Her features had softened, and her smile was calming.

"So, what about that whale?" I asked. "Where did he come from?"

"Well," she began, "About four or five summers ago, I was sitting right here, early summer, May perhaps, evening time. Out in the bay I saw an adult Kakawin circling the water in an unusual way, all alone. And then, after a while, I saw two of them; a little one, a baby, must have just been born. They stayed around for awhile, but in the morning, they were gone. I thought it was unusual to see just two of them. Usually Kakawin travel as an extended family. You rarely see them alone. I'm quite sure I saw the birth of the baby, because that's what orcas do when they give birth — they circle. Now, the scientists say Tsuxiit was swimming around with his uncle when they first noticed him. Why would he be swimming around with his uncle? But that's what the scientists are saying."

"Have you seen Tsuxiit up here at all?" I asked.

"Yes," she said, beginning to smile. "He often hangs out across the inlet, in Mooyah Bay. That's where there's the best salmon fishing."

My visit seemed to end too quickly as the *U'chuck*'s horn called me back. I said goodbye to my friend and watched the Big House grow small as the boat pulled away. The captain made an announcement: "Ladies and gentlemen, looks like Luna's gonna pay us a visit, starboard side. Don't be alarmed. He knows the vessel. He just wants to say hello. We won't slow down for him."

A black and white dart shot toward the ship from the direction of Mooyah Bay. Blacker than black, whiter than white, Luna surfaced through rainbow spray. He coasted on our wake, keeping perfect time with the speed of the ship, surfing inches from the hull of the vessel, rolling 360s in the water, showing off. He swam at great speed on his back, surfacing shiny and sleek, looking directly at us, blowing whale breath into the ship's hold. The boat listed as a hundred passengers leaned over the starboard railing. Children laughed and waved to the whale. Some adults wept with joy.

Tsuxiit stayed with our vessel for close to an hour. He was talking to us, making contact, or so it seemed to me. I was overcome with a feeling approaching bliss. My understanding of Tsuxiit was that he carried a message of purity, beauty and peace as he portrayed harmony with everything and everyone around him. I felt as one with Luna on that day.

As we approached the town of Gold River, he breached backwards and shot out toward Friendly Cove and the open sea, as if he knew the danger of being too close — the holding pen remained submerged in the inlet. The *U'chuck*'s passengers fell silent. People embraced each other with tears in their eyes. Luna was free and doing well in his adopted home.

❧

On March 10, 2006, Tsuxiit communicated with humans for the last time. He approached an idling tugboat in the waters off Gold River to accept a treat tossed to him by the crew. We shall never know why he came too close to the propeller. The Moachaht celebrated Tsuxiit's brief earthly manifestation at Yuquot on June 21, Summer Solstice. His voice can still be heard on the Internet, gentle, melodious whale song expressing to us in a timeless language the joy of being free.

A Collection

– CATHERINE LEBREDT –

WOLVES AT DUSK

Eelgrass-covered mudflats glow lime-green in the last rays of an autumn sun. Gradually, colour fades to gunmetal grey as the sun slips from sight behind a silhouette of trees. A lone heron, oblivious to the cacophony of a raucous band of passing crows, engages in its slow-motion ritual at the water's edge. At the moment of twilight, the world is wrapped in a soft cocoon of silence. Out of the melded shades of greens and greys, two dark forms appear, outlined in the afterglow. Wolves — their lean shapes unmistakable!

How long have they been there, completely exposed on the open flats? One is lying down, the other standing close by, erect and alert, its proud head turned to look in my direction: sensing a human presence, assessing the danger. In one fluid movement, both wolves are on their feet, bounding across the mudflats in wild abandon. Rearing up on their hind legs they face each other, neck-bite, then roll and bound again, one in playful pursuit of the other. I am spellbound as I watch this scene not meant for human eyes. For a few fleeting moments, I have a tiny window on the secret lives of wolves. Their lank forms, barely discernible in the fading light, pause at the water's edge. The wolves stand motionless, all senses searching the deepening dusk. One, then the other, enters the water and they swim across to darkness on the other side.

[EDS. The following excerpts are taken from her column "Up the Inlet," published for many years in *The Sound.*]

JANUARY 28 TO FEBRUARY 10, 1994

I've grown accustomed to sleeping with the loft window wide open, allowing the night sounds of the wild to penetrate my sleep and enter my dreams. And when a full moon illuminates the sky, casting its ghostly glow upon the land and sea, the night comes alive with an energy — that primal power of the moon that pulls at the lifeblood of all earthly beings. Often on moonlit nights I am roused from sleep by eagles calling, their shrill voices piercing the stillness. Great blue herons leap, squawking in terror from their treetop roosts. Sometimes they flee in vain, for the sharp eyes of the avian predators miss nothing on a night like this.

At high tide the harbour seals gather on the haul-out logs along the shore and the sounds of their moonlight revelry fill the air. It is said that harbour seals, once weaned, are silent, but it is often at night that they choose to speak. Their vocal exchanges are an eerie combination of snorts, grunts, growls and weird guttural sounds that send chills down the spines of the uninitiated. These vocalizations are accompanied by wild, boisterous play — loud splashing, flipper-slapping and occasionally, full-body breaches that end with a startling smack. It is a wonder anyone can sleep on such a night. But eventually I drift off as the sounds recede to a place in my subconscious mind. And from somewhere in my dreams I hear the deep, rhythmic hooting of a great horned owl.

FEBRUARY 11 TO FEBRUARY 24, 1994

Our initiation into inlet life began on a beautiful spring morning nearly eighteen years ago. Disillusioned with the city and longing to escape its clutches for awhile, my partner Mike and I prepared to set off on a four-month adventure into the wilds of Tofino Inlet. We had just launched our little plywood boat at the Grice Bay boat ramp and were busy stowing supplies, when an elderly couple approached us with unconcealed curiosity.

They inquired about our plans and were somewhat surprised when we disclosed our destination — Kennedy Cove. Perhaps our lack of experience and ill-preparedness was obvious to the old-timers but they didn't let on. In fact, a glimmer of recognition seemed to cross their faces. We chatted for awhile — about fishing, mostly. Then the old man asked with a detectable twinkle in his eye, if we'd ever tried bear meat at this time of year. Laughing at our inability to conceal our distaste, they both assured us it was delicious. Soon — with provisions organized to our satisfaction — we set off. The old folks wished us good luck and stood watching as we puttered away from the boat ramp.

I had the strong feeling that they were allies in our adventure — that they understood what we were doing — that perhaps, they'd been there too at one time. It was another year or two before we met Bill and Nina again — on the Crab Dock, in Tofino. And again we got to talking. By this time, we had established our little homestead at "the Point" in the mudflats of Browning Pass. Oh yes, Bill and Nina knew the place well. Sharp's Point was what the old-timers called it. A far-off look came into the old man's eyes. "Nina and I came to that point to homestead in 1919, when we were still teenagers. But that was a long time ago now." And as the old couple pulled away from the dock in their little red and white clinker, it was Mike and I who stood watching — until their boat disappeared from view — up the inlet.

FEBRUARY 25 TO MARCH 11, 1994

Perhaps it is the very essence of wilderness that we are most in danger of losing from Clayoquot Sound. The qualities intrinsic to wilderness — unspoiled natural beauty, a variety and abundance of wildlife, the absence of intrusive, mechanical noise, unbroken peace — are harder to find these days. It amazes me that here in Browning Pass after nearly two decades we still have some semblance of those qualities. But then again, as a friend has often suggested, maybe we only have the illusion that those qualities still exist.

On first glance, our inlet paradise appears much as it did when we arrived. This inlet is still a beautiful place to live and most of the time it is peaceful, sometimes unbelievably so. But the illusion of living in a wild place has walls that are gradually growing thin. A few years ago, when logging began just around the corner from us, behind Stout's Island, my illusion-bubble was burst most dramatically.

Early one morning we were awakened by the sounds of heavy machinery, fallers' chainsaws, whistles and human voices shouting over loudspeakers. It sounded like a war zone and in fact it was — part of the ongoing war on wilderness. The logging activity continued for several weeks. It was a terrible time for me. I was sure our peaceful life here had been irreparably shattered. Stout's Island is a place I once loved for its serenity. I often went there to enjoy the gardens and to appreciate the love that went into that place. I never go there now. The destruction is too real.

The intrusive sounds of logging are gone and the peace has returned, more or less. Vehicles can be heard barrelling down the road — a road that didn't exist before the logging show. And the odd gunshot penetrates the quiet. Roads, after all, bring people. And people bring their toys and their noise. A resident eagle pair, whose tree was destroyed by shake-cutters seven years ago, rebuilt their nest on the island behind our house and have successfully raised many young there. The presence of these birds is a great comfort to me. They have been here longer than we have and they survive in spite of human disturbance. These eagles play a key role in keeping my illusion intact. But I worry about them. Unlike us they cannot move on when the wilderness is gone. They will simply perish at the hands of human "progress." Their presence here serves as a gauge for me — a gauge of the strength of my illusion of wilderness. When the eagles' wilderness is eroded to the point where they can no longer survive, I'll know it's really over.

There are times when the fragility of this beautiful place really hits me in the face. During those crystal clear summer evenings, when it is so lovely we just have to sit outside; when the ethereal songs of the Swainson's thrush floats down from the forest canopy — I can feel the gentle power of this place. Until once again the illusion is broken by a mechanical drone that begins to intrude on the quiet from some far-off place, then gradually builds over several minutes to an unbearable crescendo. This is the latest assault on my illusion of wilderness — the ultralight aircraft now doing inlet tours. Some days the ultralight noise never leaves our range of hearing. It is a constant source of irritation. Once you have lived in a natural setting, largely free of intrusive, man-made noise, nothing less is satisfactory.

Perhaps the day is not far off when the experience of true wilderness, of real peace, will be impossible to attain here, or anywhere. It is a disturbing thought and a real possibility. But for now I'll hang on for dear life to the threads that hold my illusion together and be grateful that I have that much. It may be all there is.

MARCH 11 TO MARCH 25, 1994

To experience nature, to feel its subtleties, requires human perceptual ability that is capable of slowness. It requires that human beings approach the experience with patience and calm.

—Jerry Mander, *In the Absence of the Sacred* (1991)

Slow down. It's a difficult thing for many people to do and not something modern society encourages. In a world driven by cars, computers, fax machines, TVs, telephones and time schedules, the human guinea pig is trapped on a perpetual tread wheel, spinning at an ever-accelerating speed. And for most, the thought of jumping off is absolutely terrifying. But some of us do make that quantum leap out of the media-induced rat race, in an attempt to rediscover those roots to the natural world, long buried by years of urban conditioning. And the task of realigning oneself with the rhythms and cycles of nature requires an unhurried approach.

Our own transition to "slow time" began during our first summer up the inlet at Cannery Bay. Stripped of the trappings of modern society, we began to perceive the world in a different way. But it has been a gradual process, with many setbacks, as we've been pulled back and forth between the nine-to-five ritual of the "working" world and the basic needs-driven reality of this other life. After nearly twenty years, we may be approaching something resembling an independent lifestyle, where the constraints of time are largely self-imposed or dictated by nature herself. Although it may be said that the tide waits for no one, in this kind of life, it is equally true that one must often be patient and wait for the tide!

In reference to our inlet life, people often ask, "What do you do up there all day?" It is always a difficult question to answer, no matter how often it's asked. For when one is free to organize time as one wishes, everything falls naturally into the continuum of daily life. Tasks cannot be segregated or compartmentalized but are chosen out of an order of priority dictated by need and subject to the limitations of weather, season and the stages of the tide. Even our daughter's home "schooling" is integrated into family life. In this context, learning can no longer be neatly packaged into subject areas but permeates every aspect of daily activity — from planting a garden, to building a boat, to watching the eagles build a nest.

Without the distractive and disruptive presence of television, ample opportunity presents itself for imaginative and creative pursuits, for enjoying the pleasures of a good book, for appreciating nature fully, without expectation — slowly, with an attitude of patience and respect. And from our life in the "slow lane," where the rhythms and cycles of nature prevail, we've all learned it's okay to sit back now and then and do a little daydreaming — while we're waiting for the tide.

APRIL 11 TO APRIL 22, 1994

My partner Mike and I had strong ties to the land. But the ocean had a magnetism that pulled us from that path, irreversibly altering the direction of our lives. The day we purchased a twenty-six-foot live-aboard named *Waltzing Matilda* we knew there was no turning back. We were destined to become water people. For the next seven years we divided our time between our inlet homestead at Sharp's Point and our funky little double-ender, *Matilda.*

Matilda was more than a place to live. She was a means of exploring the inlets and coastline of Clayoquot Sound. Aboard *Matilda,* we came to know this area intimately, as we poked her nose into virtually every nook of the sound. We travelled in all seasons and in all kinds of weather, and out of these travels we developed a profound respect for the sea and a deep appreciation of this remarkable landscape.

Our eventual decision to build a floathouse was a natural outgrowth of these early experiences and an opportunity to embark on a new experiment — the blending of two lifestyles into one that incorporated the desirable qualities of both. Although we loved the simplicity of living on a boat, the freedom that came from having few possessions and freedom to move, we recognized the limitations. A floathouse seemed to offer the best of both worlds — the option of mobility as a safeguard (with some restrictions, of course) and the opportunity to settle in with long-term plans and projects. With some ingenuity, we could grow a garden, keep a few chickens, have a functional, well-equipped workshop and a house that was truly our own — in other words, the means to acquire some sense of independence and belonging.

Practically everything we have acquired over the past seven years has been painstakingly constructed from hand-sawn lumber and salvaged materials. This has been a painfully slow process, requiring a great deal of patience and endless hard work. But we have been rewarded in the end with a place that is very much of our own making. And each new project that we have undertaken, whether the workshop, the chicken coop or additional garden boxes, has added a new dimension to living afloat. We have often been asked why we don't make more use of the land, at least to grow a garden, for the land is close and accessible. But there's satisfaction that comes from being completely self-contained. Perhaps there's a certain amount of stubbornness involved — a desire to stay true to the life. For once you send ties out to the land, some of that independence and mobility is lost. The decision to live independent of the land requires a willingness to live within the confines of that floating space. Herein lies the challenge and beauty of living afloat. And there's a comfort in believing you can pull anchor when the time is right, leaving nothing behind — not even your footprints.

DECEMBER 2 TO DECEMBER 16, 1994
WITH THE SPIRIT OF THE SEALS

Sometimes a wild animal can touch the human spirit in a profound way, cutting through the deep layers of human conditioning that set us apart, exposing the very essence of our own animal nature and drawing us fully into the moment. No words are necessary — no explanations required.

In late September, after four nightmarish days in city hospitals where I learned I had a rare, life-threatening illness, I returned with my family to our inlet home, still too numb to feel the depth of my fear, pain and confusion. It was 5:30 in the morning when we finally pulled up to the floathouse, physically and emotionally exhausted from the experience of the previous few days. Moments after we crawled into bed, I heard Tut-Sup, the seal we'd raised from a newborn pup, flop up onto the deck. Ignoring my fatigue for the moment, I jumped out of bed and into my clothes. Tut-Sup ("sea urchin") had been left largely to his own devices for several days, and I was anxious to see how he'd fared.

Tut-Sup was obviously excited to see me. He rolled over on his back and wriggled about in an expression of pure joy. I didn't even have a fish to give him, but it didn't seem to matter. His joy at being reunited spilled over, filling me with an inexpressible feeling of relief and happiness. I knew then I'd made the right choice in returning home to begin the healing process in my own way, surrounded by the wildness of the inlet and the creatures I love.

For several days after our return, Tut-Sup stayed very close, sometimes sleeping on the back porch at night, right up against the door. I had a strong feeling he was aware of my pain and that somehow he wanted to help — to offer comfort in his own calm and quiet way.

By early October, I'd had time to put my life, my illness and my priorities into a better perspective. Surrounded by the love and concern of my family and many caring friends, I was drawn out of that dark and fearful place into the light of hope. I think Tut-Sup sensed this change, for he began to resume his old patterns, taking off for a few days at a time to fish and explore.

During one of his four-day absences in October, our lives were touched briefly by another little harbour seal. She had been picked up by fish farmers in a sorry state — one eye missing and so emaciated it was hard to believe she could still be alive. Not only was she alive, she was feisty and full of spirit. That spirit shone brightly in her one shiny black orb of an eye. For two days we fed and cared for her, but open lesions on her skin became a concern. There was a risk that I might be exposed to some harmful infection, so alternative arrangements for her care were quickly made. Rod Palm agreed to take the little seal to the Vancouver Aquarium — the only place that would accept her. But the stress of the trip was too much for her and she died on the ferry.

I was saddened by the news of the death of that little creature. But at the same time, I felt she'd come into my life during those vulnerable early days of my illness for a purpose. She showed me that a strong spirit can face great suffering, even death.

A week or so later, a third harbour seal came briefly into my life and the memory of our meeting still fills me with delight. It was during one of those blissfully beautiful October afternoons — sunny and warm without a breath of wind. I decided to take the canoe and paddle over to a small island close to home, just to be alone and soak up the sun and the silence. Later in the day, as I was preparing to leave, I noticed a seal swimming leisurely down the slough in my direction. Thinking little more about it, I launched the canoe and headed off. Within moments I heard the strong exhalations of a surfacing seal. I saw her nose, and then her full head emerge about a foot off my starboard bow. She stared directly into my eyes with an intensity I'll never forget. She soon disappeared, only to pop up again, this time off my port stern. She was so close I could have reached out and stroked her mottled grey head. I stopped paddling, letting the canoe drift with the tide. She stayed right with me, surfacing again and again beside the canoe.

Then she did something extraordinary and quite unexpected. With a burst of speed she shot under the canoe, then surfaced just ahead of me, flipped over and stuck her tail straight up out of the water. It appeared for a split second that she was standing on her head. Then with a loud "whack" her tail hit the water and she sped away. Within seconds she was back within feet of the canoe, repeating this amusing stunt. I have often seen seals play with each other in such a manner, but on this occasion I had a distinct feeling she was performing for me — somehow trying to engage me in her good-humoured sport. For at least half an hour she stayed with me, splashing and playing, popping up now and then within inches of the canoe to look steadily into my eyes. I was spellbound by her joyful display and her desire to communicate. She had me laughing out loud at her crazy antics.

By this time the tide was gaining strength, sweeping me off course. So I reluctantly took up the paddle and continued on my way. To my surprise, this seal stayed right there, swimming beside the canoe until I reached the floathouse dock. Then she surfaced one last time as if bidding goodbye, and disappeared. I'm still not sure what to make of this magical encounter. It was such a joyful, spontaneous contact, coming at a time when I needed it most. Again, this seal seemed to be telling me something — be joyful, be aware, be fully alive — here, in this moment. Uncannily, seals have had a powerful presence in my life these past few months. And each has brought a message of great importance, delivered with unmistakable clarity, on a level requiring no words. Somehow we have crossed that biological barrier that separates us as species. We have met and understood each other through the universal language of the spirit.

CONTRIBUTOR BIOGRAPHIES

GREG BLANCHETTE was born in the middle, raised in the east, and now resides in the west of Canada. After twenty years in Vancouver nursing mutually exclusive callings as engineer and adventure traveller, Greg found himself lured to the small town of Ucluelet by the Pacific Ocean. His published work includes adventure-travel magazine articles, essays, short stories, chapbooks and a miscellany of online pieces. Greg is nurturing a gift for performance poetry, and to supplement his technological roots he is thinking of going back to school to study something truly useful, like liberal arts. In the meantime he pursues an interest in Zen Buddhism.

KATE BRAID has written three prose biographies (including *Emily Carr: Rebel Artist,* with XYZ Publishing) and three prize-winning books of poetry including *To this Cedar Fountain,* a book inspired by Carr's paintings. She won the Pat Lowther poetry prize for *Covering Rough Ground* and the Vancity Book Prize for *Inward to the Bones: Georgia O'Keeffe's Journey with Emily Carr.* Braid's most recent publication, the anthology *In Fine Form: The Canadian Book of Form Poetry,* was co-edited with Sandy Shreve. In summer 2008, a second book about her experiences in construction work, *Turning Left at the Ladies,* will be published by Palimpsest Press. She lives and writes in Burnaby, B.C., where she hangs around with large trees as often as possible.

BRIAN BRETT was born in Vancouver and studied literature at Simon Fraser University. He is the author of twelve books, including *Uproar's Your Only Music* (2004) and the forthcoming *Wind River Elegies.* In 1970, along with fellow student Alan Safarik, he founded Blackfish Press, which published the periodical *Blackfish.* His journalism has appeared in almost every major newspaper in Canada. Brian inaugurated the B.C. Poetry-in-the-Schools program, introducing children in schools to world poetry. He teaches workshops and has been an active member of many writing organizations. In May 2005, Brian became the chair of The Writer's Union of Canada. Brian currently lives on a farm with his family on Salt Spring Island, B.C., where he cultivates his garden and creates ceramic forms.

HELEN CLAY hails from the hill-farming country of southwest England, where nature and wildlife were her deepest loves and sources of renewal. Family ties first brought her to Vancouver, and Clayoquot Sound insisted that she stay. Helen is following her heart as a writer and editor, currently studying at Vancouver's Douglas College. In Tofino, her articles and poems have appeared in magazines, at art festivals and on local radio. There, too, she performed in the popular amateur production of the *Vagina Monologues.* She loves kayaking and spiritual exploration — sometimes simultaneously — and is a recreational sax player.

NADINE CROOKES (KLIIAHTAH) is a member of the Ahousaht First Nation whose traditional territory encompasses the heart of Clayoquot Sound. She works for Parks Canada as Acting Superintendent at Pacific Rim National Park Reserve. She is currently working on her Master of Arts in Leadership from Royal Roads University. Nadine lives in Ucluelet, B.C., with her husband and three-year-old son. They are anxiously awaiting the arrival of the newest member of their family.

MICHAEL CURNES is the author of *Val,* a novel based on the life of Rudolph Valentino. He was a contributing writer for Tofino's *The Sound* magazine as well as the *Globe and Mail,* and has written a number of novels and screenplays, including a musical about an ancient tree that an entire village works together to save (based on Tofino's Eik Tree).

DARCY DOBELL is a science writer whose work appears in textbooks and educational resources across Canada. She also helped to found the Raincoast Education Society, which delivers environmental education programs throughout Clayoquot Sound. Darcy recently had to leave Tofino with her family, but they will be back.

KEVEN DREWS lives in a cabin at Chesterman Beach. He grew up in Surrey, B.C., but has spent almost every summer in Tofino since the late 1970s. Since 1997, he has worked as a reporter and editor at weekly and daily newspapers in British Columbia and Washington. He has also hosted local radio shows and has freelanced for a national newspaper and wire service. Keven is a cancer survivor and now runs the news site www.westcoaster.ca.

ELI ENNS, BA, JSC, is a Tla-o-qui-aht political scientist who specializes in Canadian constitutional law, international dispute resolution, and the comprehensive land claims process in British Columbia. Eli is the great grandson of Now-waas-suum (the late Harold Charlie), who was the historian and public speaker for Wickaninnish — head chief of the Tla-o-qui-aht confederation. Over the past three years Eli has worked on Tla-o-qui-aht projects, including land development at Long Beach on the Esowista Peninsula, a language revitalization initiative, and the development of the Tla-o-qui-aht Tribal Parks organization. Eli was the first Tla-o-qui-aht representative on the UNESCO Clayoquot Biosphere Trust board of directors from the fall of 2005 to the fall of 2006. Presently, Eli serves on the board of directors for several organizations both locally and nationally, including the Tonquin Foundation, Friends of Clayoquot Sound, the Tla-o-qui-aht First Nations Economic Development Corporation and the Indigenous Cooperative on the Environment. Additionally, Eli invests his time in the Clayoquot Sound Technical Planning Committee as the appointed member by the Tla-o-qui-aht hereditary chiefs.

BONNY GLAMBECK has been a professional kayak guide since 1995 and is a guide trainer living in Tofino, B.C. She is an avid expedition paddler who has explored much of the B.C. coast by kayak. Since 1988 she has been active in the movement to protect Canada's rainforest.

FRANK HARPER moved to Clayoquot in 1971. He founded, published and edited *The Sound* newspaper during the '90s, and was the travel writer for its magazine-successor. Frank wrote several three-act plays — including *Cougar Annie* — which were staged at Tofino's Community Theatre. *Journeys*, his memoirs, came out soon before his death at Catface in 2007.

KEITH HARRISON, born in Vancouver, has written half a dozen books, including *Eyemouth*, *Crossing the Gulf*, and the non-fiction novel, *Furry Creek*. He teaches at Malaspina University-College and lives on Hornby Island. His most recent novel is *Elliot & Me*.

DIANNE IGNACE spent her first twenty-one years on the Saskatchewan prairies, fourteen of those in Meota, a resort town even smaller than Tofino. After finishing college in 1973, Dianne moved from Saskatoon to Winnipeg and worked in the kitchen at the University of Manitoba. Not impressed by her brief taste of city life, she moved to Campbell River, Vancouver Island, to spend time with her older brother and his wife. Soon after, while working in Tofino, Dianne met her husband and joined him at Hesquiat in 1975.

SUSAN KAMMERZELL is most happy with her existence when she is somewhere in the mountains on foot, although bikes, snowshoes and canoes also have their uses. She also gets a kick out of making her personal footprint smaller each year, and is fascinated by the project of helping herself and other humans to claim their sense of joyful connection with the biosphere we inhabit. She is a fine listener, with a healer's heart, and has a determined mind and a passion for both her planet and her species. The quotation used in this book came from Susan's article "Homebody: No Other Heaven."

BETTY KRAWCZYK is a fearless activist and great-grandmother who wrote *Clayoquot: The Sound of My Heart* and *Lock Me Up or Let Me Go.* She has raised eight children, run in federal elections for the Green party, spent much of her recent life behind bars for blockading logging roads and highways, and is now running for mayor in Vancouver.

VALERIE LANGER has been engaged in forest conservation campaigns for twenty years, with a brief stint in an Italian circus. In 2006 she joined Forest Ethics' Great Bear Rainforest campaign after having worked as the Forest Campaign Director for the Friends of Clayoquot Sound from 1988 to 2003. She pioneered the market campaign strategy of changing logging companies' practices by engaging with their commercial customer base abroad. Valerie helped found two other environmental organizations in which she is still engaged. Her idea of a good time is thinking about what will induce a change in the nature of humans' interactions with their biosphere. Hiking for days is also very good. As a freelance writer she has written feature articles and reports on the topics of forestry and First Nations for the *Globe and Mail, Paper Europe* and for national environmental organizations. Valerie has worked professionally and voluntarily teaching adult literacy. She holds a degree in Semiotics from the University of Toronto.

CATHERINE LEBREDT was born in Winnipeg. After moving to Clayoquot Sound as a young woman with her husband Mike, she became influenced by her surroundings in a way that led her to begin writing. Here she had their daughter, April, and wrote a column for *The Sound* called "Up the Inlet," about life in a floathome. Catherine died in 1995 at the age of forty-two.

ROB LIBOIRON was born in Medicine Hat, Alberta, and raised in B.C.'s Okanagan Valley. Living remotely in Clayoquot Sound for some years, Rob is gifted with a natural talent for the fine arts, enjoying years of tattooing, painting and, more recently, carving. Long interested in writing, Rob is just now venturing into that field.

HANNE LORE is a west coast multimedia artist and filmmaker whose visual work has been exhibited internationally since the 1970s. She has contributed to many public celebrations: Habitat, the Edinburgh International Festival, the Vancouver International Children's Festival, the Asia Pacific Festival, Expo '86, the Austin Texas Festival of Light, and the Hong Kong New Year Celebration. Hanne has created documentary films, written film scripts, illustrated children's books and published several short stories. She lived in Bamfield from 1970 to 1990 and now resides in Vancouver.

CHRISTINE LOWTHER is the author of *New Power* and *A Cabin in Clayoquot.* Her work has been featured on CBC Radio's North by Northwest and published in anthologies and periodicals, including the *Vancouver Sun, The Fiddlehead, The Beaver, Tofino Time* and *The Sound.* She saves trees when necessary. Chris loves punk rock, dancing, her kayak and her bicycle. She lives in Clayoquot Sound.

ADRIENNE MASON is a writer and editor living in Tofino, B.C. She writes for both children and adults, with her books focusing on science, nature and west coast history. Currently, Adrienne is the managing editor of *KNOW,* a science magazine for children. Her collection of heart-shaped stones (and shells and wood and beach glass) continues to grow as she explores the west coast with her friends and family.

JANIS MCDOUGALL is a part-time member of the Clayoquot Writers' Group and has appeared in the anthology *Salt in Our Blood.* She has lived in Clayoquot for thirty years.

SHERRY MERK spent the best decade of her life in Clayoquot Sound, during which she also enlivened the Clayoquot Writers' Group, where she is greatly missed. She is presently living and writing in Port Alberni, and hopes one day to return to Clayoquot.

ALEXANDRA MORTON is a renowned biologist, photographer, artist and writer. She is well known for her slide shows, films, television appearances and books, which include *Listening to Whales, Siwiti: A Whale's Story, In the Company of Whales*, and *Heart of the Raincoast* (co-written with Billy Proctor). Her research has appeared in *Proceeds of the National Academy of the Sciences of the United States of America (PNAS), North American Journal of Fisheries Management, Alaska Fishery Research Bulletin, Marine Mammal Science, Transactions of the American Fisheries Society, Canadian Journal of Fisheries and Aquatic Sciences, ICES Journal of Marine Science,* and *Canadian Journal of Zoology.* She says of her home in the Broughton Archipelago, "It is my place on the planet."

SUSAN MUSGRAVE lives on Vancouver Island and on the Queen Charlotte Islands/Haida Gwaii. Her most recent book is *You're in Canada Now . . . : A Memoir of Sorts,* published by Thistledown in 2005. *When the World is Not Our Home: Selected Poems 1985–2000* will be published in 2009. She teaches Creative Writing online in the Optional Residency MFA program at the University of British Columbia.

BRIONY PENN's nature column for *Monday* magazine became the book *A Year on the Wild Side.* Living on Salt Spring Island with her family, she is a daredevil activist setting an excellent example for her two sons.

DAVID PITT-BROOKE wrote *Chasing Clayoquot: A Wilderness Almanac.* He practised veterinary medicine for a decade and a half, with digressions into wildlife research that included breeding falcons, collaring caribou and implanting radios in rattlesnakes. From 1987 to 1995, he was an environmental education officer for Parks Canada. In 1995 he established his own communications business and has written on a wide range of topics, from grizzly bears to critical path analysis. In June 2002, he received a Canadian Science Writers' Association Award for "Outstanding Contribution to Science Journalism in Canadian Media."

CAROLYN REDL is a naturalist, writer and teacher who collects humorous narratives of people's unusual interactions with wildlife. She also has a yen for the Arctic, satisfied through travels north of 60° and studies of northern women's stories.

JOANNA ROBINSON is a PhD student in Sociology at the University of British Columbia. Her research interests include social movements, the environment, globalization and labour. Currently she is completing a cross-national comparative study of communities that organize against the privatization and commodification of water. Previously, she completed her BA in Sociology and a post-graduate diploma in environment at McGill University and worked for several years in the environmental non-profit community in B.C. She currently serves as vice-president of the board of directors at the Society Promoting Environmental Conservation (SPEC) in Vancouver, where she is active in campaigns to promote food safety and security and solid waste recycling.

ANITA SINNER is completing her PhD at the University of British Columbia. She teaches in art education, and her research interests focus on arts-based educational research, life writing and new media. She is a photographic artist living in Sooke, B.C.

JOANNA STREETLY is the author of the novel *Silent Inlet* and a book of nonfiction, *Paddling Through Time.* Most recently she has written for *Between Interruptions,* an anthology edited by Cori Howard. Joanna was a regular contributor to *BBC Wildlife Magazine* and past editor of *The Sound.* She also edited *Salt in Our Blood,* an anthology of west coast moments. These days Joanna's home floats near Strawberry Island, Clayoquot Sound.

ANDREW STRUTHERS grew up in Scotland, Uganda, British Columbia and Japan. His account of the Clayoquot protests was published as *The Green Shadow* and won the National Magazine Award for Humour. Andrew wrote *The Last Voyage of the Loch Ryan* (2004) in Victoria, where he lives with his daughter Pasheabel and works as a filmmaker and writer.

DAVID B. TINDALL is an Associate Professor in the Department of Sociology at the University of British Columbia. He has conducted a variety of interconnected studies examining the environmental movement and environmental concern amongst the general public in B.C., and in Canada more widely. His studies have considered the importance of social and environmental values, socio-demographic factors, and social networks for explaining concern with environmental issues and participation in pro-environmental behaviours. He has also studied media coverage of the conflict over ancient forests in B.C. Much of his current research is focused on the social dimensions of climate change.

CHANDRA WONG is a writer, photographer and artist with training in biology and teaching. Her experience includes writing for community newspapers and magazines, and most recently, a historical publication about the Fort St. John hospital. She currently lives on the north side of the Peace River, and looks forward to what comes of her current venture into creative nonfiction.